Pitman Research Notes in Mathematics Series

Submission of proposals for consideration
Suggestions for publication, in the form of outlines and representative samples, are invited by the Editorial Board for assessment. Intending authors should approach one of the main editors or another member of the Editorial Board, citing the relevant AMS subject classifications. Alternatively, outlines may be sent directly to the publisher's offices. Refereeing is by members of the board and other mathematical authorities in the topic concerned, throughout the world.

Preparation of accepted manuscripts
On acceptance of a proposal, the publisher will supply full instructions for the preparation of manuscripts in a form suitable for direct photo-lithographic reproduction. Specially printed grid sheets can be provided and a contribution is offered by the publisher towards the cost of typing. Word processor output, subject to the publisher's approval, is also acceptable.

Illustrations should be prepared by the authors, ready for direct reproduction without further improvement. The use of hand-drawn symbols should be avoided wherever possible, in order to maintain maximum clarity of the text.

The publisher will be pleased to give any guidance necessary during the preparation of a typescript, and will be happy to answer any queries.

Important note
In order to avoid later retyping, intending authors are strongly urged not to begin final preparation of a typescript before receiving the publisher's guidelines. In this way it is hoped to preserve the uniform appearance of the series.

Addison Wesley Longman Ltd
Edinburgh Gate
Harlow, Essex, CM20 2JE
UK
(Telephone (0) 1279 623623)

Titles in this series. A full list is available from the publisher on request.

350 Progress in elliptic and parabolic partial differential equations
A Alvino, P Buonocore, V Ferone, E Giarrusso, S Matarasso, R Toscano and G Trombetti
351 Integral representations for spatial models of mathematical physics
V V Kravchenko and M V Shapiro
352 Dynamics of nonlinear waves in dissipative systems: reduction, bifurcation and stability
G Dangelmayr, B Fiedler, K Kirchgässner and A Mielke
353 Singularities of solutions of second order quasilinear equations
L Véron
354 Mathematical theory in fluid mechanics
G P Galdi, J Málek and J Necas
355 Eigenfunction expansions, operator algebras and symmetric spaces
R M Kauffman
356 Lectures on bifurcations, dynamics and symmetry
M Field
357 Noncoercive variational problems and related results
D Goeleven
358 Generalised optimal stopping problems and financial markets
D Wong
359 Topics in pseudo-differential operators
S Zaidman
360 The Dirichlet problem for the Laplacian in bounded and unbounded domains
C G Simader and H Sohr
361 Direct and inverse electromagnetic scattering
A H Serbest and S R Cloude
362 International conference on dynamical systems
F Ledrappier, J Lewowicz and S Newhouse
363 Free boundary problems, theory and applications
M Niezgódka and P Strzelecki
364 Backward stochastic differential equations
N El Karoui and L Mazliak
365 Topological and variational methods for nonlinear boundary value problems
P Drábek
366 Complex analysis and geometry
V Ancona, E Ballico, R M Mirò-Roig and A Silva

Pavel Drábek (Editor)
University of West Bohemia, Czech Republic

Topological and variational methods for nonlinear boundary value problems

20th Seminar in Partial Differential Equations

LONGMAN

Addison Wesley Longman Limited
Edinburgh Gate, Harlow
Essex CM20 2JE, England
and Associated Companies throughout the world.

Published in the United States of America
by Addison Wesley Longman Inc.

First published 1997

AMS Subject Classifications (Main) 35J65, 35L70, 47H11
(Subsidiary) 47H15, 49J35, 47H12

ISSN 0269-3674

ISBN 0 582 30921 2

British Library Cataloguing in Publication Data

A catalogue record for this book is
available from the British Library

Printed and bound in Great Britain
by Biddles Ltd, Guildford and King's Lynn

Contents

Preface

The **20th Seminar in Partial Differential Equations** took place from June 5–9, 1995 in Cholín (on Slapy Dam, 40 km south of Prague). It was organized by the Department of Mathematics of the University of West Bohemia in Pilsen and by the Society of Czech Mathematicians and Physicists. The Seminar continues a long tradition started twenty years ago. Founded in 1976 by Svatopluk Fučík and Alois Kufner, it has been held regularly every year and starting in 1978, every 4th year it has been transformed into an International Spring School "Nonlinear Analysis, Function Spaces and Applications." On the occasion of this anniversary we present below the list of speakers and the topics which have been presented since 1976. Since 1992 the official language has been English but the main goal of the Seminar has not changed: to present a complete survey of some particular recent topics in PDEs to a wider mathematical audience, including nonspecialists.

1st Seminar 1976 Popelná (South-East Bohemia)
S. Fučík and A. Kufner: The methods of solution of nonlinear differential equations

2nd Seminar 1977 Pernink (North-East Bohemia)
J. Stará and O. John: Nonlinear differential equations
B. Opic: Degenerated differential equations

3nd Seminar = 1st International Spring School 1978 Horní Bradlo (East Bohemia, Proceedings published by Teubner – Texte zur Mathematik, Leipzig, 1979, Editors: S. Fučík and A. Kufner)
J. Camus: On a class of weighted Sobolev spaces
Y. A. Dubinskij: Some problems of the theory of Sobolev spaces of infinite order and of nonlinear equations
D. E. Edmunds: Embeddings of Sobolev spaces
J. – P. Gossez: Orlicz–Sobolev spaces and nonlinear elliptic boundary value problems
K. Gröger: Initial value problems for elastoplastic and elasto–viscoplastic systems
I. Hlaváček: Some variational methods for nonlinear mechanics
H. Triebel: Recent developments in the theory of function spaces and linear regular elliptic differential equations
E. Zeidler: Lectures on Lyusternik–Schmirelman theory for indefinite nonlinear eigenvalue problems and its applications
A. Avantaggiati: Compact imbedding theorems in Sobolev weight spaces (not included in Proceedings)

4th Seminar 1979 Příhrazy (East Bohemia)
J. Stará and O. John: Selected part from the bifurcation theory and applications
S. Míka: Finite elements method

5th Seminar 1980 Alšovice (North Bohemia)
M. Kučera: Variational inequalities and the problem of optimal control
A. Kufner: Weighted Sobolev spaces–properties and applications
P. Drábek: Weak solutions of superlinear elliptic boundary value problem

6th Seminar 1981 Manětín (West Bohemia)
Z. Vlášek: Numerical solutions of plain potential flows
J. Franců: Homogenization
J. Stará: $G-$ and $\Gamma-$convergence
J. Franců a J. Stará: Historical remarks on the homogenization and $G-$ a $\Gamma-$convergence
P. Doktor: Variational and quasivariational inequalities

7th Seminar = International Spring School 1982 Písek (South Bohemia, Proceedings published by Teubner - Texte zur Mathematik, Leipzig, 1982, Editors: O. John and A. Kufner)
V. I. Burenkov: Mollifying operators with variable step and their application to approximation by infinitely differentiable functions
M. Giaquinta: On differentiability of the extremals of variational integrals
K. P. Hadeler: Nonlinear differential equations from biology
P. L. Lions: Fully nonlinear elliptic equations and applications
V. G. Maz'ya: Theory of multipliers in spaces of differentiable functions and its applications
I. V. Skrypnik: Topological methods of investigation of operator equations and nonlinear boundary value problems
J. R. L. Webb: Approximation solvability of nonlinear equations
T. M. Rassias: The influence of topology to nonlinear analysis
C. Baiocchi: Free boundary value problems and variational inequalities (not included in Proceedings)
J. Ball: Calculus of variations and nonlinear elastostatics (not included in Proceedings)

8th Seminar 1983 Doksy (North Bohemia)
Š. Schwabik: Fundaments of optimal control theory
S. Míka, P. Drábek, J. Voldřich, V. Kohout, A. Kufner: Stokes problem

9th Seminar 1984 Velká Úpa (Giant Mountains)
J. Kačur: An application of the Rothe method in mathematical physics
I. Straškraba: Systems of quasilinear hyperbolic equations
J. Reif: Convex analysis

10th Seminar 1985 Malá Lučivná (West Slovakia)
J. Haslinger: Numerical solution of contact problems
M. Štědrý: On some inverse problem

11th Seminar = 3rd International Spring School 1986 Litomyšl (East Bohemia, Proceedings published by Teubner–Texte zur Mathematik, Leipzig, 1986, Editors: M. Krbec, A. Kufner and J. Rákosník)
L. I. Hedberg: Nonlinear potential theory and Sobolev spaces
H. P. Heinig: Weighted estimates for classical operators
R. Johnson: A survey on tent spaces and their application to weighted inequalities
A. G. R. McIntosh: Clifford algebras and the double–layer potential operator
J. Musielak: Some applications of generalized Orlicz spaces in approximation theory and Fourier series
J. Peetre: Generalizations of Hankel operators
B. Ruf: Multiplicity results for nonlinear elliptic equations
O. V. Besov: Integral representation of functions and imbedding theorems for domains with the flexible horn property
L. D. Kudryavtsev: On stabilization of functions and free boundary variational problems on unbounded intervals
D. R. Adams: The classification problem for Besov and Triebel–Lizorkin capacities (not included in Proceedings)
H. Lange: Nonlinear Schrödinger equation (not included in Proceedings)

12th Seminar 1987 Lipovce (East Slovakia)
P. Krejčí: The methods of solution of hysteresis equations
B. Opic: Hardy–type inequality and imbedding theorems for weighted Sobolev spaces

13th Seminar 1988 Podhradí nad Dyjí (South Moravia)
M. Fila and J. Filo: Qualitative analysis of some nonlinear diffusion problems
A. Kufner: Solvability of elliptic problems on nonsmooth domains

14th Seminar 1989 Hřensko (North Bohemia)
P. Drábek: Solvability and bifurcations of nonlinear equations
V. Šverák: Nonlinear equations and the weak convergence

15th Seminar = 4th International Spring School 1990 Roudnice nad Labem (near Prague, Proceedings published by Teubner - Texte zur Mathematik, Leipzig, 1990, Editors: M. Krbec, A. Kufner, B. Opic and J. Rákosník)
J. Garcia - Cuerva: Factorization of operators and weighted norm inequalities
H. P. Heinig: Weighted inequalities in Fourier analysis
V. Kokilashvili: Weighted estimates for classical integral operators
V. Mustonen: Mappings of monotone type: Theory and applications
L. E. Persson: Generalizations of some classical inequalities and their applications

G. F. Roach: Aspects of nonlinear scattering theory
C. G. Simader: The weak Dirichlet and Neumann problem for the Laplacian in L^q for bounded and exterior domains. Applications.
S. K. Vodop'yanov: Boundary behaviour of differentiable functions and related topic

16th Seminar 1991 Cheb (West Bohemia)
E. Feireisl: Atractors for non–autonomous hyperbolic equations
J. Kottas: Regularity of solutions to variational inequalities

17th Seminar 1992 Cheb (West Bohemia)
J. Král: Removable singularities of solutions of partial differential equations
J. Appell: Fixed point theory and nonlinear boundary value problems
W. Sickel: Superposition of functions in Bessel potential spaces

18th Seminar 1993 Vyškov (Central Moravia)
I. V. Skrypnik: Topological characteristics of fully nonlinear parabolic problems
H. Leinfelder: The Schrödinger equation

19th Seminar = 5th International Spring School 1994 Praha (Proceedings published by Prometheus Publishing House, Praha 1994, Editors: M. Krbec, A. Kufner, B. Opic and J. Rákosník)
F. Chiarenza: L^p-regularity for systems of PDE's, with coefficients in VMO
D. E. Edmunds: Recent developments concerning entropy and approximation numbers
B. Kawohl: On the shape of solutions to some variational problems
F. J. Martín–Reyes: Weights, one–sided operators, singular integrals and ergodic theorems
V. D. Stepanov: Weighted norm inequalities for integral operators and related topics
G. Talenti: Inequalities in rearrangement invariant function spaces
R. L. Wheeden: Poincaré–Sobolev and isoperimetric inequalities, maximal functions, and half–space estimates for the gradient
E. T. Sawyer: On the oblique derivative problem (not included in Proceedings).

20th Seminar 1995 Cholín (near Prague)
J. Berkovits: Some extensions of topological degree theory with applications to nonlinear problems
J. – P. Gossez: Some remarks on the antimaximum principle
S. I. Pohozaev: The fibering method in nonlinear variational problems
A. Szulkin: Index theories for indefinite functionals and applications

21st Seminar 1996 Železná Ruda (West Bohemia)
P. Drábek and A. Kufner: Nonlinear degenerated and singular equations
T. Runst and W. Sickel: Sobolev spaces of fractional order, Nemytzkij operators and nonlinear partial differential equations.

The Proceedings of the International Spring Schools were published in English while proceedings of Seminars # 4, 5, 6, 8, 9, 10, 12, 13, 14, and 16 were published in Czech by the Technical University in Pilsen. Proceedings of Seminars # 17 and 18 were published in English by the University of West Bohemia in Pilsen. It should also be mentioned that the lectures of I. V. Skrypnik given at 18th Seminar are not included in Proceedings of that seminar. Since Prof. Skrypnik has finally delivered the manuscript to the Editor, we decided to publish his contribution in this volume.

During the period 1976 – 1987 the organizing committee of the Seminar had been chaired by Alois Kufner, assisted by phenomenal organizer Jiří Holenda whose role in the Seminar cannot be estimated highly enough. Since 1988 the Editor of this volume has been responsible for the program, assisted by Petr Dolanský, Jana Lepičová and Marta Míková.

Let us also emphasize another important feature of our Seminars and International Spring Schools: they are usually held at an attractive place, outside of the big cities, giving the participants an opportunity both to work hard and to relax in beautiful natural surroundings. Due to this particular atmosphere our Seminar has been the starting point of several international collaborations which have culminated in joint research projects, scientific papers and monographs. Another important aspect is the fact that during the Cold War our Seminar and the International Spring Schools provided a bridge between "Eastern" and "Western" Mathematics, and contributed to the exchange of ideas and opinions among mathematicians from all over the world. Even before 1989, the "Iron Curtain" never existed at our conferences.

We would like to finish this preface by introducing the authors of contributions to this volume. Juha Berkovits is professor at the University of Oulu, Finland and he works on the topological degree and its applications to nonlinear boundary value problems. Jean–Pierre Gossez is professor at the University of Brussels (Université Libre), Belgium and since the late sixties he has been working in various directions of nonlinear analysis. Professor Stanislav Ivanovich Pohozaev works at the Steklov Mathematical Institute of the Russian Academy of Sciences in Moscow, Russia. His work consists of an enormous number of deep results, the most famous being the so–called Pohozaev's identity discovered in the mid sixties. Andrzej Szulkin is professor at the University of Stockholm, Sweden. His recent research is devoted to the study of critical point theory and its applications to nonlinear equations and Hamiltonian systems. Professor Igor Vladimirovich Skrypnik works at the Institute of Applied Mathematics and Mechanics of the Ukrainian Academy of Sciences in Donetsk, Ukraine. His research is devoted mainly to the theory of nonlinear partial differential equations of higher order.

I would like to thank all the authors for their contributions, Addison Wesley Longman Ltd. for publishing this volume, and Ing. Josef Daněk and Jana Lepičová from the University of West Bohemia in Pilsen for preparation of the camera-ready manuscript. I would also like to thank my colleague and collaborator Professor Alois Kufner from the Mathematical Institute of the Academy of Sciences in Prague for his help and support during the organization of our Seminars and the preparation of this volume. Finally I would like to express my gratitude to the Grant Agency of the Czech Republic for the grant #201/94/0008 which provided financial help for the organization of the 20th Seminar in PDEs.

September 1996 The Editor

J BERKOVITS

Some extensions of topological degree theory with applications to nonlinear problems

1. Introduction

The problems in mathematics and mathematical physics can typically be reduced to a study of the set of solutions for an equation $F(u) = h$, where F is a map between appropriate spaces X and Y, and h is an element of Y. A useful way to obtain information about the solution set is topological degree theory. The topological degree is mainly used in the study of ordinary and partial differential equations, integral equations, bifurcation theory and in proving fixed point theorems. Degree theory also provides a natural technique for proving "mapping theorems" including generalizations of the Jordan curve theorem.

The notion of the topological degree was first introduced by L.E.J.Brouwer in 1912 [Brou] for continuous mappings in finite dimensional spaces. The classical Brouwer degree in $\mathbb{R}^n$ can be defined either by using algebraic topology or analytically (see e.g.[Cro] and [Llo]). Leray and Schauder generalized in 1934 [Le-Sch] the degree theory for compact perturbations of identity in infinite dimensional Banach spaces. Their construction is based on the Brouwer degree.

Since 1934, various generalizations of degree theory have been introduced (see e.g. [Llo],[Sch]). However, it was not until in 1972 and 1973 that Fuhrer [Fu] and Amann and Weiss [Am-We] proved the Brouwer degree to be uniquely determined by only a few basic conditions. These conditions provide a natural basis for the formal definition of a classical topological degree (see [Brow], [Ber1] and [Llo]).

In 1983 F.E.Browder [Brow] extended the concept of the topological degree to nonlinear mappings of monotone type from a real reflexive Banach space into its dual space. His method is based on Galerkin approximations for which the finite dimensional Brouwer degree is well-defined. Another contruction which is based on so-called "elliptic super-regularization" and Leray-Schauder degree was given in [Be-Mu1] and [Ber1]. Mappings of monotone type are widely used in the study of partial elliptic differential equations of general divergence form (see [Be-Mu2], and references therein).

In [Be-Mu3] the Leray-Schauder degree is uniquely extended to a class of nonlinear mappings on a real Hilbert space. It can be used to treat effectively problems connected to hyperbolic equations, e.g. semilinear wave equations, see [Be-Mu4],[Be-Mu5],[Be-Mu6],[Be-Mu8],[Be-Mu9],[Be-Mu10][Ber3], [Be-Be1] and [Be-Be2]. This extension is closely related to the multiple valued concidence degree given by Mawhin in 1972 [Ma1], [Ma2]. A degree theory applicable to certain parabolic problems is constructed in [Be-Mu7]. It is based on the degree theory for mappings of monotone type, where the monotonicity conditions are given with respect to the graph norm topology induced by some maximal monotone map. In applications the maximal monotone part is typically the time derivative.

Some recent contributions to degree theory can be found in [Kit] and [Tie]. In the first one there is a construction of a degree theory for multivalued mappings, which can be applied to partial differential equations with strongly nonlinear perturbations. The degree theory presented in the second paper gives a possibility to consider mappings between non-reflexive spaces e.g. differential operators in Orlicz-Sobolev spaces. Further extensions of the degree theory for multivalued mappings are given in [Be-Ti], where applications to partial differential equations with discontinuities are considered.

The present paper is organized as follows. In Section 2 we give the formal definition of classical topological degree and comment briefly the nature of different conditions. The third section is devoted to a construction of topological degree – we shall present three different extensions of degree theory which can be applied to different type of problems. In our fourth section we illuminate the basic use of degree theory in applications, that is, how to prove existence, surjectivity and multiplicity results by using homotopy argument and "change of degree"-principle. For simplicity we shall consider only one specific problem – one dimensional wave-equation with different kind of nonlinearities. We shall close this paper by a short discussion about the Jordan separation theorem in our final Section 5.

2. Formal definition of the degree

In this section we shall study the general structure of topological degree theory. We first recall the definition of classical topological degree.

Definition 2.1: Let X and Y be topological spaces and let $\mathcal{O}$ be a family of open subsets of X. For each G in $\mathcal{O}$, we associate a class $\mathcal{F}_G$ of maps of $\overline{G}$ into Y.For any map $F \in \mathcal{F}_G, G \in \mathcal{O}$, and for any $h \in Y$ outside $F(\partial G)$ we associate an integer $d(F, G, h)$.

The integer-valued function d is said to be a classical topological degree if the following conditions are satisfied.

(a) If $d(F, G, h) \neq 0$, then there exists a solution for the equation $F(u) = h$ in G.

(b) (Additivity) If $G_0 \in \mathcal{O}$ is a subset of $G \in \mathcal{O}$ and $F \in \mathcal{F}_G$, then the restriction $F|_{\overline{G}_0}$ belongs to $\mathcal{F}_{G_0}$. Let G_1 and G_2 be a pair of disjoint subsets of G belonging to $\mathcal{O}$ such that $h \notin F(\overline{G} \setminus (G_1 \cup G_2))$. Then

$$d(F, G, h) = d(F, G_1, h) + d(F, G_2, h).$$

(c) (Invariance under homotopy) For a suitably defined family $\mathcal{H}_G, G \in \mathcal{O}$, of admissible homotopies $\{F_t : 0 \leq t \leq 1\}$ from $\overline{G}$ into Y and for any continuous curve

$\{h_t : 0 \leq t \leq 1\}$ in Y with $h_t \notin F_t(\partial G)$ for all $t \in [0,1]$, the value $d(F_t, G, h_t)$ is constant in t on $[0,1]$.

(d) (Normalizing map) There exists an injective map J from X into Y such that $J|_{\overline{G}}$ lies in $\mathcal{F}_G$ for each G in $\mathcal{O}$, and if $h \in J(G)$, then $d(J, G, h) = +1$.

Let us choose $X = Y = \mathbb{R}^n, \mathcal{O}$ all open bounded sets, $\mathcal{F}_G$ all continuous maps from $\overline{G}$ into $\mathbb{R}^n$, $\mathcal{H}_G$ all continuous homotopies, i.e., continuous maps from $[0,1] \times \overline{G}$ into $\mathbb{R}^n$, and $J = I$ the identity map. Then we get the classical degree theory derived by Brouwer in 1912 [Brou]. Moreover, the Brouwer degree d_B is uniquely determined by conditions (a)-(d) listed above.

Similarly, the unique Leray-Schauder d_{LS} is obtained by choosing $X = Y$ a real Banach space, $\mathcal{O}$ all open bounded sets, $\mathcal{F}_G$ all continuous maps F of $\overline{G}$ into X such that $I - F$ is compact, $\mathcal{H}_G$ the family of continuous homotopies of the form $I - T$, where $T : [0,1] \times \overline{G} \to X$ is compact, and again $J = I$ the identity map. Due to the uniqueness the Brouwer and Leray-Schauder degrees coincide whenever $X = \mathbb{R}^n$.

We shall now study more closely the conditions imposed on a degree function in Definition 2.1. From now on we assume that X and Y are normed linear spaces and $\mathcal{O}$ consists of all open bounded sets of X. Suppose that $\mathcal{H}_G$ contains at least *affine homotopies*, i.e., homotopies of the form $(1-t)F_0 + tF_1, 0 \leq t \leq 1$, where F_0 and F_1 belong to $\mathcal{F}_G$. Now $\mathcal{F}_G \subset \mathcal{H}_G$, since any $F \in \mathcal{F}_G$ can be written as $F = (1-t)F + tF$. We assume that for any homotopy $F_t, 0 \leq t \leq 1$, in $\mathcal{H}_G$ and for any continuous curve $h_t, 0 \leq t \leq 1$, in Y also $F_t - h_t$ determines an admissible homotopy. Especially, $F - h \in \mathcal{F}_G$ whenever $F \in \mathcal{F}_G$ and $h \in Y$. Let us further assume that each $F \in \mathcal{F}_G$ takes closed sets of $\overline{G}$ into closed sets of Y. In particular, the sets $F(\partial G)$ and $F(\overline{G})$ are closed. It should be noted that all these assumtions hold in the context of Brouwer and Leray-Schauder degree theories and they hold also for all degree theories mentioned in this paper.

The following three basic results follow directly from the general definition.

Theorem 2.1. *(Boundary dependence) Suppose that $F \in \mathcal{F}_G$ and $T \in \mathcal{F}_G$ coincide on the boundary of G and $h \in Y \backslash F(\partial G)$. Then $d(F, G, h) = d(T, G, h)$.*

Proof. The result is a direct consequence of the homotopy invariance property applied to the affine homotopy $F_t := (1-t)F + tT$, since $F_t(u) = F(u) \neq h$ for all $u \in \partial G$ and $t \in [0,1]$. □

Theorem 2.2. *Let $G \in \mathcal{O}$ and $F \in \mathcal{F}_G$ be fixed. Then $d(F, G, \cdot)$ is constant on each open component of the open set $Y \backslash F(\partial G)$.*

Proof. Let Δ be an open component of the open set $Y \backslash F(\partial G)$, that is, a maximal set of points which can be connected by a continuous curve in $Y \backslash F(\partial G)$. Let h and f be any two fixed elements of Δ. Taking the homotopy $F_t \equiv F$ and using the homotopy invariance one sees that $d(F, G, h) = d(F, G, f)$ completing the proof. □

In view of the previous theorem we can define the value of the degree with respect to the component Δ by setting

$$d(F, G, \Delta) = d(F, G, h),$$

where $h \in \Delta$ is arbitrary. Theorem 2.2 shows that some condition like $h \notin F(\partial G)$ cannot be avoided in the definition of a degree, since otherwise $d(F, G, h)$ would be constant for all h in Y, which contradicts the normalizing condition (d) of Definition 2.1.

Theorem 2.3. *Let $F \in \mathcal{F}_G$ and $h \in Y \backslash F(\partial G)$ be given. Then $d(F, G, h) = d(F - h, G, 0)$.*

Proof. Consider the affine homotopy $F_t = (1-t)F + t(F-h)$ and a continuous curve $h_t = (1-t)h$, where $0 \leq t \leq 1$. Clearly $F_t(u) \neq h_t$ for all $u \in \partial G$ and $t \in [0,1]$. Hence $d(F_t, G, h_t)$ is constant in t on $[0,1]$ and the values $t = 0$ and $t = 1$ give the desired result. □

It is useful to examine to what extent the conditions from (a) to (d) in Definition 2.1 are independent. It is easily seen that (a) follows from the additivity property (b). Indeed, if we first choose $G = G_1 = G_2 = \emptyset$ and apply (b), then $d(F, \emptyset, h) = d(F, \emptyset, h) + d(F, \emptyset, h)$, which shows that $d(F, \emptyset, h) = 0$. If now $F \in \mathcal{F}_G$ and $h \notin F(\overline{G})$, then by choosing $G_1 = G_2 = \emptyset$ and applying (b) again we get $d(F, G, h) = 0$, which is exactly condition (a). However, the condition (a) is so characteristic property of any degree theory that it is presented as a separate condition.

Let us now give some examples which will clarify the significance of conditions (b), (c) and (d).

Example 2.1 Let d be a degree function and define a new integer valued function d_1 by setting

$$d_1(F, G, h) = [d(F, G, h)]^2$$

whenever the righthand side is well-defined. Then d_1 satisfies conditions (a),(c) and (d) but the additivity property (b) does not hold. Hence (b) is not a consequence of the other conditions.

Example 2.2 Define d_2 by setting $d_2(F, G, h) = 0$ for all $h \notin F(\partial G)$. All conditions except (d), the normalizing condition, are clearly satified. Thus (d) is independent of the other conditions (a), (b) and (c).

Example 2.3 Since the normalizing map J is injective we can define

$$d_3(F, G, h) = \begin{cases} +\ 1, & \text{if } h \in J(G) \text{ and } F(J^{-1}(h)) = h \\ \ \ 0 & \text{otherwise} \end{cases}$$

for any $h \notin F(\partial G)$. It is not hard to see that d_3 satisfies (a), (b) and (d). For any $h \in J(G), h \neq 0$ we have $d_3(J, G, h) = +1$, since $J(J^{-1}(h)) = h$, but $d_3(J - h, G, 0) = 0$, since $(J-h)(J^{-1}(0)) = -h \neq 0$. Thus (c), invariance under homotopy cannot be valid in view of Theorem 2.3.

It should be noted that once a degree function is constructed for some class of mappings in such a way that (a), (b) and (d) hold, then it is often possible to choose the class of homotopies so narrow that also (c) is satisfied. However, the practical value of a degree theory depends on the homotopies available. Thus it important to pay attention to the choice of admissible homotopies – at least affine homotopies should be included. The basic existence and surjectivity results are obtained using affine homotopies. Instead of the normalizing map we shall consider a more general notion of a reference map. By a *reference map* we mean admissible injection $T \in \mathcal{F}_X$ for which $d(T, G, w) \neq 0$ whenever $G \subset X$ is a open bounded set and $w \in T(G)$. We get

Theorem 2.4. *(Continuation theorem) Assume that $T : X \to Y$ is a reference map, G is an open bounded set in X and $w \in T(G)$. If $F \in \mathcal{F}_G$ and*

$$(1-t)F(u) + tT(u) \neq (1-t)h + tw \quad \text{for all } u \in \partial G \text{ and } t \in [0,1],$$

then the equation $F(u) = h$ admits at least one solution u in G.

We shall close this section by a fixed point theorem, which includes the Brouwer and Schauder fixed point theorems as special cases.

Theorem 2.5. *(Fixed point theorem) Assume that $X = Y$, the normalizing map is I and $F : \overline{B}_R(0) \to \overline{B}_R(0)$ is mapping such that $I - F \in \mathcal{F}_{B_R(0)}$, where $B_R(0) = \{u \in X \mid \|u\| < R\}$. Then F has a fixed point, i.e., there exists $u \in \overline{B}_R(0)$ such that $F(u) = u$.*

Proof. If $u - F(u) = 0$ for some $u \in \partial B_R(0)$ the proof is complete. Hence we can assume that $u - F(u) \neq 0$ for all $u \in \partial B_R(0)$. We consider the affine homotopy $F_t = I - tF, 0 \leq t \leq 1$. If $F_t(u) = 0$ for some $t \in [0,1]$ and $u \in \partial B_R(0)$, then

$$R = \|u\| = t\|F(u)\| \leq tR$$

implying $t = 1$ and thus $u - F(u) = 0$ on the boundary giving a contradiction. Hence by the homotopy argument $d(I - F, B_R(0), 0) = d(I, B_R(0), 0) = +1$ and the conclusion follows by property (a). □

3. On the construction of degree

In this section we shall consider the construction of a degree function. In general, the definition of a degree theory for some new class of mappings requires suitable approximation cheme, that is, any map can be approximated by maps for which the degree theory already exists. To illuminate this we shall study three different situations.

A. Elliptic case. Let X be a real reflexive separable Banach space, renormed if necessary such that X and its dual space X^* are locally uniformly convex [Tro]. We consider mappings acting from X into X^*. For simplicity we shall assume that all mappings are defined on the whole of X. The norm convergence is denoted by $\to$ and the weak convergence by $\rightharpoonup$ and the continuous pairing between X and X^* by $\langle \cdot, \cdot \rangle$. We recall that a mapping $F : X \to X^*$ is

- *bounded*, if it takes any bounded set into a bounded set,
- *demicontinuous*, if $u_k \to u$ implies $F(u_k) \rightharpoonup F(u)$,
- *monotone*, if $\langle F(u) - F(v), u - v \rangle \geq 0$ for all $u, v \in X$,
- *of class* (S_+), if for any sequence with $u_k \rightharpoonup u$ and $\limsup \langle F(u_k), u_k - u \rangle \leq 0$, it follows that $u_k \to u$.

We shall show how to construct a degree theory for bounded demicontinuous maps of class (S_+). In 1983 F.Browder gave a contruction based on finite dimensional Galerkin-approximations and Brouwer degree [Brow]. However, we present another construction given in [Ber1], which uses "elliptic super- regularization" and Leray-Schauder degree. First we recall the following embedding theorem due to Browder and Ton [Br-To].

Theorem 3.1. *Let X be a reflexive separable Banach space. Then there exist a separable Hilbert space H and a compact linear injection $\psi : H \to X$ such that $\psi(H)$ is dense in X.*

We define a further map $\hat{\psi} : X^* \to H$ by setting

$$(\hat{\psi}(w), v) = \langle w, \psi(v) \rangle \quad \text{for all } v \in H \text{ and } w \in X^*,$$

where $(\cdot, \cdot)$ stands for the inner product of H. It is easily seen that also $\hat{\psi}$ is a compact linear injection.

Let G be an open bounded set of X and $F : \overline{G} \to X^*$ a bounded demicontinuous mapping of class (S_+). We associate to F a family of mappings defined by

$$F_\lambda = I + \lambda \psi \hat{\psi} F \qquad \text{for any } \lambda > 0.$$

For any fixed $\lambda > 0, F_\lambda$ maps $\overline{G}$ into X. Moreover, F_λ is of the type $I+$ compact, a map for which the Leray- Schauder degree is defined. The basic result in view of our construction is the following

Lemma 3.1. *Let $F : \overline{G} \to X^*$ be a bounded demicontinuous mapping of the class (S_+) and assume that $0 \notin F(\partial G)$. Then there exists $\lambda_0 > 0$ such that $0 \notin F_\lambda(\partial G)$ and $d_{LS}(F_\lambda, G, 0)$ is constant in λ for all $\lambda > \lambda_0$.*

Proof. Assume that the first assertion is false. Then we can find sequences $\{\lambda_k\}$, $\lambda_k \to \infty$, $\{u_k\} \subset \partial G$ such that $F_{\lambda_k}(u_k) = 0$. At least for subsequences we can write

$$u_k \rightharpoonup u \in X, \quad F(u_k) \rightharpoonup w \in X^*.$$

Clearly $\psi\hat{\psi}F(u_k) = -\lambda_k^{-1}u_k \to 0 = \psi\hat{\psi}w$ and hence $w = 0$. Thus

$$\begin{aligned}\limsup\langle F(u_k), u_k - u\rangle &= \limsup\langle F(u_k), u_k\rangle \\ &= \limsup\langle F(u_k), -\lambda_k\psi\hat{\psi}F(u_k)\rangle \\ &= \limsup\{-\lambda_k\|\hat{\psi}F(u_k)\|_H^2\} \leq 0,\end{aligned}$$

where $\|\cdot\|_H$ stands for the norm of H. Consequently, $u_k \to u \in \partial G$ and $F(u_k) \rightharpoonup F(u) = w = 0$ giving a contradiction. Hence there exists $\lambda_0 > 0$ such that $F_\lambda(u) \neq 0$ for all $u \in \partial G$ and $\lambda > \lambda_0$. As a consequence, $d_{LS}(F_\lambda, G, 0)$ is well-defined for all $\lambda > \lambda_0$. The second assertion follows trivially by the homotopy invariance property of d_{LS}, since $F_\lambda, \lambda_1 \leq \lambda \leq \lambda_2$, defines an admissible homotopy in Leray-Schauder theory. □

By the preceding lemma it is relevant to define a new function d by setting

$$d(F, G, 0) = \lim_{\lambda\to\infty} d_{LS}(F_\lambda, G, 0),$$

whenever $0 \notin F(\partial G)$. For any $h \in X^*\backslash F(\partial G)$ we set

$$d(F, G, h) = d(F - h, G, 0).$$

It can be shown that the integer valued function d given above really satifies the conditions (a) to (d) of definition 2.1 [Ber1]. In this case the normalizing map will be the *duality map* J, which is uniquely determined by the conditions

$$\|J(u)\| = \|u\|, \quad \langle J(u), u\rangle = \|u\|^2 \quad \text{for all } u \in X.$$

The duality map is strictly monotone continuous bijection of class (S_+). Note that J is linear if and only if X is Hilbert space and in that case J^{-1} is the Frechet-Riesz linear isometry.

We say that $F_t : \overline{G} \to X^*, 0 \leq t \leq 1$, is *a bounded homotopy of class* (S_+), if F_t is uniformly bounded in t and if conditions

$$u_k \rightharpoonup u, t_k \to t \quad \text{and } \limsup\langle F_{t_k}(u_k), u_k - u\rangle \leq 0,$$

imply $u_k \to u$ and $F_{t_k}(u_k) \rightharpoonup F_t(u)$. The class of admissible homotopies also contains affine homotopies between (S_+)-maps.

The degree theory is unique and it can be extended to all demicontinuous mappings of class (S_+), not only to bounded ones as above, see [Ber1].

The main field of applications of the degree theory for mappings of class (S_+) lies in the study of nonlinear elliptic partial differential operators in divergence form

$$Au(x) = \sum_{|\alpha|\leq m}(-1)^{|\alpha|}D^\alpha A_\alpha(x, u(x), \ldots, D^m u(x)), \quad x \in \Omega,$$

where Ω is an open subset in $\mathbb{R}^N$. If the functions A_α satisfy certain natural growth, monotonicity and coercivity conditions, then the corresponding operator between appropriate Sobolev spaces will be pseudomonotone or of class (S_+), see [Be-Mu2] and the references therein.

B. Parabolic case. Let X be a real reflexive Banach space. We recall that a mapping $T : D(T) \subset X \to X^*$ is called *maximal monotone*, if it is monotone and its graph

$$G(T) = \{(u, T(u)) \in X \times X^* | u \in D(T)\}$$

is maximal as a monotone set in $X \times X^*$, that is, if $(u, w) \in X \times X^*$ is such that $\langle w - y, u - v\rangle \geq 0$ for all $(v, y) \in G(T)$, then $u \in D(T)$ and $w = T(u)$. Using the duality map it is possible characterize the maximal monotony. Indeed, a map T is maximal monotone if and only if the range of $T + \lambda J$ is the whole space X^* for every $\lambda > 0$, see [De]. Let $L : D(L) \subset X \to X^*$ be a fixed densely defined linear closed maximal monotone map. Since the graph of L is a closed set in $X \times X^*$, the space $Y := D(L)$ equipped with the graph norm

$$\|u\|_Y = \|u\|_X + \|Lu\|_{X^*}, \quad u \in Y,$$

becomes a real reflexive Banach space. We assume that both X and Y with corresponding dual spaces are locally uniformly convex. We shall deal with bounded demicontinuous maps from X into X^*, which are monotone type with respect to the graph norm topology. For instance, a map $F : X \to X^*$ is said to be of *class* (S_+) *with respect to* $D(L)$, if the conditions

$$\{u_k\} \subset D(L), u_k \rightharpoonup u, Lu_k \rightharpoonup Lu, \limsup\langle F(u_k), u_k - u\rangle \leq 0$$

imply that $u_k \to u$ in X. Note that this class is wider than the usual class (S_+). Similarly one can define other classes of mappings of monotone type with respect to $D(L)$.

We shall study equations of form

$$\text{(E)} \qquad Lu + F(u) = h, \quad u \in D(L),$$

where F is of class (S_+) with respect to $D(L)$ and $h \in X^*$. Let j stand for the natural embedding of Y to X and j^* its adjoint from X^* to Y^*. For any open bounded set G of X and any bounded demicontinuous map $F : \overline{G} \to X^*$ of class (S_+) with respect to $D(L)$ we denote

$$\hat{F} = j^* \circ F \circ j : j^{-1}(\overline{G}) \to Y^*.$$

Similarly we denote $\hat{L} = j^* \circ L \circ j$. In what follows we also need an auxiliary map $P : Y \to Y^*$ defined by

$$(P(u), v) = \langle Lv, J^{-1}(Lu)\rangle, \qquad u, v \in Y,$$

where $(\cdot,\cdot)$ denotes the pairing between Y and Y^* and J^{-1} is the inverse of the duality map $J : X \to X^*$. Note that for all those $u \in Y$ for which $P(u) \in j^*(X^*)$, we have $J^{-1}(Lu) \in D(L^*)$ and hence

$$P(u) = j^* L^* J^{-1}(Lu).$$

For each map $L + F$ we associate a family of mappings defined by

$$\hat{F}_\epsilon = \hat{L} + \hat{F} + \epsilon P, \quad \epsilon > 0.$$

Let h in X^* be given. We denote

$$S = \{u \in j^{-1}(\overline{G}) | \hat{F}_\epsilon(u) = j^* h \quad \text{for some } \epsilon > 0\}.$$

The relationship between $L + F$ and its approximation $\hat{F}_\epsilon$ is shown in the following

Lemma 3.2. *Let $G \subset X$ be an open bounded set and $F : \overline{G} \to X^*$ a bounded demicontinuous map of class (S_+) with respect to $D(L)$ and $h \in X^*$. Then*

(a) $\hat{F}_\epsilon$ is a bounded demicontinuous map of class (S_+) from $j^{-1}(\overline{G}) \subset Y$ to Y^*.

(b) There exists $R > 0$ such that $S \subset B_R(Y) = \{v \in Y | \|v\|_Y < R\}$.

(c) If $h \notin (L + F)(\partial G \cap D(L))$, then there exists $\epsilon_0 > 0$ such that

$$j^* h \notin \hat{F}_\epsilon(j^{-1}(\partial G)) \quad \text{for all } 0 < \epsilon < \epsilon_0.$$

Proof. To prove (a) assume that $\{u_k\} \subset \overline{G} \cap D(L)$ with $u_k \rightharpoonup u$ in Y and $\limsup(\hat{F}_\epsilon(u_k), u_k - u) \leq 0$. Then $u_k \rightharpoonup u$ in X, $Lu_k \rightharpoonup Lu$ in X^* and

$$\limsup\{\langle Lu_k - Lu, u_k - u\rangle + \langle F(u_k), u_k - u\rangle + \epsilon\langle Lu_k - Lu, J^{-1}(Lu_k) - J^{-1}(Lu)\rangle\}.$$

Since L is monotone and J^{-1} is strictly monotone we can conclude that $\limsup\langle F(u_k), u_k - u\rangle \leq 0$. Hence $u_k \to u$ in X and also

$$\lim\langle Lu_k - Lu, J^{-1}(Lu_k) - J^{-1}(Lu)\rangle = 0$$

implying by the properties of J^{-1} that $Lu_k \to Lu$ in X^*. Thus $u_k \to u$ in Y and the proof of assertion (a) is complete.

Clearly the set S is bounded in X. Assertion (b) states that S is bounded in Y, too. Without loss of generality we can assume that $h = 0$. Let $u \in S$ be arbitrary. Then for some $\epsilon > 0$

$$\langle Lu, v\rangle + \langle F(u), v\rangle + \epsilon\langle L^* J^{-1}(Lu), v\rangle = 0$$

for all $v \in D(L)$. Note that $J^{-1}(Lu) \in D(L^*)$ since $P(u) \in j^*(X^*)$. By the density of $D(L)$ in X the above equation hold for all $v \in X$. Hence we can insert $v = J^{-1}(Lu)$ and use the known properties of J and L^* to get

$$\|Lu\|_{X^*}^2 \leq \|F(u)\|_{X^*}\|Lu\|_{X^*}.$$

Thus there exists a constant $R > 0$ such that $\|u\|_Y < R$ for all $u \in S$ and the value of R is independent of ϵ.

To prove assertion (c) we shall argue by contradiction. Assume again that $h = 0$. Let us assume that there exist sequences $\{\epsilon_k\}$ and $\{u_k\} \subset j^{-1}(\partial G)$ such that $\epsilon_k \to 0+$ and

$$\hat{L}u_k + \hat{F}(u_k) + \epsilon_k P(u_k) = 0.$$

By the previous part (b) of this theorem the sequence $\{u_k\}$ is bounded in Y and hence we can write $u_k \rightharpoonup u \in D(L)$ in X and $Lu_k \rightharpoonup Lu$ in X^*, at least for a subsequence. Using the basic properties of L and J^{-1} we get

$$\limsup\langle F(u_k), u_k - u\rangle = \limsup\{-\langle Lu_k - Lu, u_k - u\rangle$$
$$- \epsilon_k\langle Lu_k - Lu, J^{-1}(Lu_k) - J^{-1}(Lu)\rangle\} \leq 0.$$

Thus $u_k \to u$ in X and consequently $\langle Lu, v\rangle + \langle F(u), v\rangle = 0$ for all $v \in D(L)$. Since $D(L)$ is dense in X we have

$$Lu + F(u) = 0, \quad \text{with } u \in \partial G \cap D(L)$$

contradicting our assumption. The proof is complete. □

In view of Lemma 3.2 we can now define the degree $d(L + F, G, h)$ by using the corresponding family $\{\hat{F}_\epsilon\}$ of (S_+)-maps. Note however that the open set $j^{-1}(G)$ is not generally bounded and

$$\overline{j^{-1}(G)} \subset j^{-1}(\overline{G}), \quad \partial(j^{-1}(G)) \subset j^{-1}(\partial G).$$

Assume that $h \notin (L + F)(\partial G \cap D(L))$. By Lemma 3.2 there exist $\epsilon_0 > 0$ and $R > 0$ such that

$$\hat{F}_\epsilon^{-1}(j^*h) \subset B_R(Y) \quad \text{and } \hat{F}_\epsilon(u) \neq j^*h \text{ for all } u \in \partial G_R(Y), 0 < \epsilon < \epsilon_0,$$

where we have denoted $G_R(Y) = j^{-1}(G) \cap B_R(Y)$. Thus the (S_+)-degree

$$d_{S_+}(\hat{L} + \hat{F} + \epsilon P, G_R(Y), j^*h)$$

is well-defined for all $0 < \epsilon < \epsilon_0$. It is easy to see that the value of degree is independent of ϵ and we can define

$$d(L + F, G, h) = \lim_{\epsilon \to 0+} d_{S_+}(\hat{F}_\epsilon, G_R(Y), j^*h).$$

The integer-valued function d given above satisfy the conditions of classical topological degree [Be-Mu7]. The normalizing map in this case will be $L + J$. The admissible homotopies are bounded homotopies of class (S_+) with respect to $D(L)$ containing again affine ones.

The degree theory contructed in this sub-section can be used to deal with nonlinear parabolic initial-boundary value problem of the type

$$\begin{cases} \dfrac{\partial u}{\partial t} + A(u) = h & \text{in } \Omega \times [0.T] \\ u(x,0) = 0 & \text{in } \Omega \\ D^\alpha u(x,t) = 0 & \text{on } \partial\Omega \times [0,T] \text{ for all } |\alpha| \leq m-1, \end{cases}$$

where Ω is an open bounded subset in $\mathbb{R}^N$, h is a given function defined in $\Omega \times [0,T]$ and A is a divergence operator of order $2m$,

$$Au(x,t) = \sum_{|\alpha| \leq 0} (-1)^{|\alpha|} D^\alpha A_\alpha(x,t,u(x,t),\ldots,D^m u(x,t))$$

satisfying some growth, monotonicity and coercivity conditions. The maximal monotone linear operator L will be the time derivative – for details see [Be-Mu7], [Be-Mu11].

C. Hyperbolic case. Let H be a real separable Hilbert space with inner product $(\cdot,\cdot)$ and corresponding norm $\|\cdot\|$. Let M be a closed subspace of H and $Q : H \to M$ the corresponding orthogonal projection. We denote $P = I - Q : H \to M^\perp$ and consider mappings of the form

$$\mathcal{F}_G = \{F = Q(I-C) + PN : \overline{G} \to H\}$$

where G is an open bounded set in H, C is compact and N is bounded demicontinuous map of the class (S_+). Note that in case $M^\perp$ is finite dimensional, the projection P is compact, and $\mathcal{F}_G$ contains exactly the Leray-Schauder type maps. If $\dim M < \infty$, the class $\mathcal{F}_G$ consists of (S_+)-type maps, only. Hence we assume that both M and $M^\perp$ are infinite dimensional. Since any Leray-Schauder type map is of class (S_+) we have $I - C = Q(I-C) + P(I-C) \in \mathcal{F}_G$, i.e., $\mathcal{F}_G$ contains always the Leray-Schauder maps.

We shall construct a degree theory for the class $\mathcal{F}_G$ of admissible mappings. It will be an extension of Leray-Schauder degree in Hilbert space setting. The construction can be done by using Galerkin approximation with respect to the space $M^\perp$, see [Be-Mu3]. However, we shall give here another construction which is based on similar argument as used in the case of mappings of class (S_+) earlier in this paper. Indeed, let $\psi : M^\perp \to M^\perp$ be some fixed compact selfadjoint injection. To any $F = Q(I-C) + PN \in \mathcal{F}_G$ we associate a family of mappings defined by

$$F_\lambda = I + QC + \lambda\psi^2 PN, \quad \text{for any } \lambda > 0.$$

Clearly F_λ is a Leray-Schauder type map. In analogy with Lemma 3.1 we have

Lemma 3.3. *Assume that $F = QI - C) + PN \in \mathcal{F}_G$ and $0 \notin F(\partial G)$. The there exists $\lambda_0 > 0$ such that $0 \notin F_\lambda(\partial G)$ and $d_{LS}(F_\lambda, G, 0)$ is constant in λ for all $\lambda > \lambda_0$.*

Proof. We shall argue by contradiction. Assume that the first assertion is not valid. Then we can find sequences $\{\lambda_k\}, \lambda_k \to \infty, \{u_k\} \subset \partial G$ such that $F_{\lambda_k}(u_k) = 0$. At least for subsequences we can write $u_k \rightharpoonup u$, $N(u_k) \rightharpoonup w$ and $C(u_k) \to c$. The equation $F_{\lambda_k}(u_k) = 0$ is equivalent to

$$\begin{cases} Qu_k - QC(u_k) = 0 \\ Pu_k + \lambda_k \psi^2 PN(u_k) = 0 \end{cases}$$

and thus $Qu_k \to Qu = Qc$ and $\psi^2 PN(u_k) \to 0 = \psi^2 Pw$ implying $w = 0$. Hence

$$\begin{aligned} &\limsup(N(u_k), u_k - u) = \limsup(PN(u_k), Pu_k - Pu) \\ &= \limsup(PN(u_k), -\lambda_k \psi^2 PN(u_k)) \\ &= \limsup\{-\lambda_k \|\psi PN(u_k)\|^2\} \leq 0, \end{aligned}$$

which implies $u_k \to u \in \partial G$ and consequently $F(u) = Q(u - C(u)) + PN(u) = 0$ giving a contradiction. The second assertion again follows from the homotopy invariance property of d_{LS}. □

It is relevant to define d by setting

$$d(F, G, 0) = \lim_{\lambda \to \infty} d_{LS}(F_\lambda, G, 0),$$

whenever $0 \notin F(\partial G)$ and $d(F, G, h) = d(F - h, G0)$, for any $h \notin F(\partial G)$. The integer-valued function d satisfies the conditions of classical topological degree. The normalizing map is $I = QI + PI$ and the admissible homotopies are of the form

$$F_t = Q(I - C_t) + PN_t, \quad 0 \leq t \leq 1,$$

where $I - C_t$ is a Leray-Schauder homotopy and N_t a bounded homotopy of class (S_+). Moreover, the degree function d is unique, in particular, its value is independent of the choice of the compact injection ψ.

In wiev of applications, we consider following situation. Let $L : D(L) \subset H \to H$ be a fixed densely defined closed linear operator such that $\operatorname{Im} L = (\operatorname{Ker} L)^\perp$. Hence $\operatorname{Im} L$ is closed and $\operatorname{Ker} L = \operatorname{Ker} L^*$. Let L_0 stand for the restriction of L into $\operatorname{Im} L \cap D(L)$ and assume that its inverse $K = L_0^{-1} : \operatorname{Im} L \to \operatorname{Im} L$ is compact. We shall study the equation

(E) $$Lu - N(u) = h, \quad u \in D(L),$$

where the nonlinearity $N : H \to H$ is assumed to be bounded demicontinuous map of class (S_+) and $h \in H$. We choose $M = \operatorname{Im} L$ and $M^\perp = \operatorname{Ker} L$. It is easy to see that (E) is equivalent to

$$Q(u - KQN(u)) + PN(u) = f, \quad u \in H,$$

where $f = (KQ - P)h$. Note that if $u \in H$ satisfies the preceding equation, then the fact that $KQ - P$ is the right inverse of $L - P$ implies $u \in D(L)$. Clearly $Q(I - KQN) + PN \in \mathcal{F}_G$ for any open bounded set G in H. In order to simplify our notations we define a further degree function " deg " by setting

$$\deg(L - N, G, h) \equiv d(Q(I - KQN) + PN, G, (KQ - P)h)$$

for any open bounded set $G \subset H$ and $h \in H$ such that $h \notin (L - N)(\partial G \cap D(L))$. Notice that the normalizing map in this setting will be $L-P$, since $(KQ-P)(L-P) = I$ on $D(L)$.

Remark. In fact it is sufficient that the map N is bounded demicontinuous and $PN = PT$ for some (S_+)-map T. This fact is used above, since P is not of the class (S_+) (because we assumed that $M, M^\perp$ are infinite dimensional), but $P = PI$, where the identity map is of class (S_+). Hence the degree for $L - P$ is well-defined.

We can now tackle hyperbolic partial differential equations. A standard example is the one dimensional wave equation with periodic-Dirichlet boundary conditions

$$\text{(WE)} \qquad \begin{cases} u_{tt} - u_{xx} - g(t, x, u) = h(t, x), & (t, x) \in \mathbb{R} \times (0, \pi), \\ u(t, 0) = u(t, \pi) = 0 & \text{for all } t \in \mathbb{R}, \\ u(\cdot, x) & \text{is } 2\pi\text{-periodic for all } x \in (0, \pi). \end{cases}$$

We shall consider the wave equation more closely in later sections. Therefore we give the exact formulation of (WE) in the space $H = L^2(\Omega)$, where $\Omega =]0, 2\pi[\times]0, \pi[$. Denote by C^2 twice continuously differentiable functions $v : \mathbb{R} \times [0, \pi] \to \mathbb{R}$ such that $v(t, 0) = v(t, \pi) = 0$ for all $t \in \mathbb{R}$ and v is 2π-periodic in t. We shall assume that the function $(t, x, s) \to g(t, x, s)$ from $\Omega \times \mathbb{R}$ to $\mathbb{R}$ is a Caratheodory function, i.e., measurable in (t, x) for each $s \in \mathbb{R}$ and continuous in s for a.a. $(t, x) \in \Omega$. Moreover, we assume that g is 2π-periodic in t and satisfies the usual growth condition

$$|g(t, x, s)| \leq c_0|s| + h_0(t, x)$$

for all $s \in \mathbb{R}$ and a.a. $(t, x) \in \Omega$ with some constant $c_0 > 0$ and $h_0 \in L^2(\Omega)$. Then g induces a bounded continuous Nemytskii operator $N : H \to H$ by

$$N(u)(t, x) = g(t, x, u(t, x))$$

for all $u \in H$ and a.a. $(t,x) \in \Omega$. A *weak solution* of (WE), where $h \in H$, is any $u \in L^2(\Omega)$ satisfying

$$(u, v_{tt} - v_{xx}) - (N(u), v) = (h, v) \quad \text{for all } v \in C^2.$$

Denoting

$$\psi_{jk} = \begin{cases} \frac{\sqrt{2}}{\pi}\sin(jx)\sin(kt), j,k \in \mathbb{Z}_+ \\ \frac{1}{\pi}\sin(jx), \qquad j \in \mathbb{Z}_+, k = 0 \\ \frac{\sqrt{2}}{\pi}\sin(jx)\cos(kt), j, -k \in \mathbb{Z}_+ \end{cases}$$

the set $\{\psi_{jk} | j \in \mathbb{Z}_+, k \in \mathbb{Z}\}$ forms an orthonormal basis in $L^2(\Omega)$ and the wave operator $\frac{\partial^2}{\partial t^2} - \frac{\partial^2}{\partial x^2}$ with periodic Dirichlet boundary conditions has in $L^2(\Omega)$ abstract realization

$$Lu = \sum_{j,k} (j^2 - k^2)(u, \psi_{jk})\psi_{jk}$$

with

$$D(L) = \{u \in L^2(\Omega) | \sum_{j,k} |j^2 - k^2|^2 |(u, \psi_{jk})|^2 < \infty\}.$$

It is not hard to see that $L : D(L) \subset H \to H$ is densely defined closed selfadjoint linear operator with closed range $\operatorname{Im} L = (\operatorname{Ker} L)^\perp$. Moreover, L has a pure point spectrum of eigenvalues

$$\sigma(L) = \{\lambda_{jk} = j^2 - k^2 | j \in \mathbb{Z}_+, k \in \mathbb{Z}\}$$

with corresponding eigenfunctions $\{\psi_{jk}\}$. Clearly $\sigma(L)$ is unbounded from above and from below, any eigenvalue $\lambda_{jk} \neq 0$ has finite multiplicity $m(\lambda_{jk})$, but $\operatorname{Ker} L$ is infinite dimensional. Thus the right partial inverse K of L from $\operatorname{Im} L$ to $\operatorname{Im} L$ is compact. It can be shown that $u \in L^2(\Omega)$ is a weak solution of (WE) if and only if

$$Lu - N(u) = h \quad \text{with } u \in D(L),$$

which is special case of the equation (E). It is easy to see that the mapping N is srongly monotone and thus also of the class (S_+), if g is strongly monotone, that is, for some constant $\mu > 0$

$$\frac{g(t,x,s_1) - g(t,x,s_2)}{s_1 - s_2} \geq \mu$$

for all $s_1 \neq s_2$ and $(t,x) \in \Omega$. Similarly, the map N is pseudomonotone, if g is nondecreasing in s.

In the sequel it will often be more convenient to renumerate the spectrum $\sigma(L)$ as a strictly increasing sequence

$$\cdots < \lambda_{-2} < \lambda_{-1} < \lambda_0 = 0 < \lambda_1 < \lambda_2 < \cdots$$

and the basis accordinly as $\{\psi_{jk} | j \in \mathbb{Z}, k = 1, \ldots m(\lambda_j)\}$. We shall use this notation further without any comment.

4. APPLICATIONS–SEMILINEAR WAVE EQUATION

In what follows we shall concentrate on results where the degree theoretic arguments seem to be essential. In general, we shall omit most of the technicalities of the proofs, but try to point out where and how the degree theory is used. We shall start with some remarks on the reference maps in the case of one dimensional wave equation. Results related to the material which is presented in this section can be found e.g. in [Am1],[Am2],[Am-Ze],[Br-Ni], [Dr-Hu],[Dr],[Hi],[La-Mc],[Ma2],[Ra].

4.1 Reference maps

The efficient use of homotopies requires always some known mapping having non-zero degree. The normalizing map is not always the best possible alternative. In order to illuminate this we consider the one dimensional wave equation prescribed in the previous section. Indeed, we can use at least three different type of reference maps. The choice will depend on the structure of the original problem. We have the following

Lemma 4.1. *For any $\lambda \in \sigma(L)$ we denote by P_λ the orthogonal projection from H to $\mathrm{Ker}(L - \lambda I)$. The mappings*

(a) $L - P$, *where $P = P_0$ is the orthogonal projection to* $\mathrm{Ker}\, L$,
(b) $L - \tau I$, *where $\tau > 0$ and $\tau \notin \sigma(L)$,*
(c) $L - \lambda I \mp P_\lambda$, *where $\lambda > 0, \lambda \in \sigma(L)$,*

are reference maps. For any open bounded set G in H we have as a property of normalizing map $\deg(L - P, G, h) = +1$ for all $h \in (L - P)(G \cap D(L))$. Moreover,

$$\deg(L - \tau I, G, h) = (-1)^{\sum_{0<\lambda_{jk}<\tau} m(\lambda_{jk})}$$

whenever $h \in (L - \tau I)(G \cap D(L))$. Similarly

$$\deg(L - \lambda I \mp P_\lambda, G, h) = \begin{cases} (-1)^{\sum_{0<\lambda_{jk}\leq\lambda} m(\lambda_{jk})} \\ (-1)^{\sum_{0<\lambda_{jk}<\lambda} m(\lambda_{jk})}, \end{cases}$$

whenever $h \in (L - \lambda I \mp P_\lambda)(G \cap D(L))$.

The proof of the preceding theorem is a direct application of Theorem 6.1 of [Ber2] together with the fact that in n-dimensional space $d(-I, G, 0) = (-1)^n$ if $0 \in G$, see [Llo]. In [Be-Mu3] one can find how the above reference maps and continuation theorem are employd in the study of surjectivity and Landesman-Lazer type resonance results. However, the aim of this paper is to tackle problems where the use of degree is more sophisticated. Anyway, we shall utilize the results of Lemma 4.1 in the following sections.

4.2 Existence of nontrivial solutions

Let $g(t,x,s)$ be a Caratheodory function from $\Omega \times \mathbb{R}$ to $\mathbb{R}$ such that $g(t,x,\cdot)$ is strictly increasing and

$$\alpha \le \frac{g(t,x,s)}{s} \le \beta, \quad \text{for all } s \neq 0, (t,x) \in \Omega,$$

where $\beta > \alpha > 0$ are constants. Then the Nemytski operator N generated by g from $H = L^2(\Omega)$ to H is bounded continuous and $N(0) = 0$. We shall study the existence of nontrivial weak solutions for the semilinear wave equation

$$\text{(SWE)} \qquad \begin{cases} u_{tt} - u_{xx} = g(t,x,u), & (t,x) \in (0,\pi) \times \mathbb{R}, \\ u(t,0) = u(t,\pi) = 0, & t \in \mathbb{R}, \\ u(\cdot,x) & \text{is } 2\pi\text{-periodic for all } x \in (0,\pi). \end{cases}$$

For simplicity we assume that $g = g(s)$, only. We are interested in the case where N interacts with the spectrum of L in the sense that a finite number of eigenvalues with finite multiplicity are crossed by the function $\frac{g(s)}{s}$, when $|s|$ goes from 0 to ∞. Indeed, denote

$$l_0 = \liminf_{s \to 0} \frac{g(s)}{s}, \quad l_\infty = \limsup_{|s| \to \infty} \frac{g(s)}{s}$$

and assume that for some positive eigenvalues $\lambda_q < \cdots < \lambda_{q+m}$ of the wave operator L we have

$$\lambda_{q-1} < \alpha \le l_\infty < \lambda_q < \cdots < \lambda_{q+m} < l_0 \le \beta < \lambda_{q+m+1}.$$

We refer to the eigenvalues $\lambda_q, \dots, \lambda_{q+m}$ as crossed eigenvalues. We define the orthogonal projections by

$$P_1 u = \sum_{j<q} (u, \psi_{jk})\psi_{jk}, P_2 u = \sum_{q \le j \le q+m} (u, \psi_{jk})\psi_{jk}, P_3 u = \sum_{j > q+m} (u, \psi_{jk})\psi_{jk},$$

for any u in H. Then $H = H_1 \oplus H_2 \oplus H_3$ with $H_i = \operatorname{Im} P_i (i = 1,2,3)$, where H_2 is finite dimensional and called crossing space. Our main result is the following.

Theorem 4.1. *If the dimension of the crossing space H_2 is odd, then the equation* (SWE) *admits at least one nontrivial solution $u \in D(L)$.*

The idea of the proof is to show that there exist $R > r > 0$ such that

$$\deg(L - N, B_R, 0) \neq \deg(L - N, B_r, 0),$$

where B_ρ denotes an open ball with radius ρ and center at the origin. Then the additivity property of degree implies

$$\deg(L - N, B_R \backslash \overline{B}_r, 0) = \deg(L - N, B_R, 0) - \deg(L - N, B_r, 0) \neq 0$$

and hence there exists a weak solution u of (SWE) with $r < \|u\| < R$. We need the following a priori estimates.

Lemma 4.2. *There exist constants $R > r > 0$ such that*

(a) $Lu - l_\infty u - \mu(N(u) - l_\infty u) \neq 0$ for all $0 \leq \mu \leq 1$ and $u \in D(L)$ with $\|u\| \geq R$.

(b) $Lu - l_0 u - \mu(N(u) - l_0 u) \neq 0$ for all $0 \leq \mu \leq 1$ and $u \in D(L)$ with $0 < \|u\| \leq r$.

Now the conclusion follows, since in view of Lemma 4.1 and the homotopy invariance property

$$\deg(L - N, B_R, 0) = \deg(L - l_\infty I, B_R, 0) = (-1)^{\sum\limits_{0<\lambda_j\leq\lambda_{q-1}} m(\lambda_j)}$$

and

$$\deg(L - N, B_r, 0) = \deg(L - l_0 I, B_r, 0) = (-1)^{\sum\limits_{0<\lambda_j\leq\lambda_{q+m}} m(\lambda_j)}$$

and $\dim H_2 = \sum_{j=q}^{q+m} m(\lambda_j)$ is odd.

The proof of Lemma 4.2 which can be found in [Be-Mu6] is based on the division $H = H_1 \oplus H_2 \oplus H_3$ together with sharp estimates of

$$(N(u), P_1 u \pm P_2 u - P_3 u) = \int\limits_\Omega g(u)(P_1 u \pm P_2 u - P_3 u)$$

taking into account the behavior of g as $|u(t,x)|$ is near zero or very large, accordingly. Also heavily used is the fact that for any $u \in D(L)$ we have

$$\begin{cases} (Lu, P_1 u) \leq \lambda_{q-1}\|P_1 u\|^2, \\ \lambda_q \|P_2\|^2 \leq (Lu, P_2 u) \leq \lambda_{q+m}\|P_2 u\|^2, \\ (Lu, P_3 u) \geq \lambda_{q+m+1}\|P_3 u\|^2. \end{cases}$$

It is interesting to notice that the monotonicity of g is not needed in proof of the priori estimates of Lemma 4.2. Hence similar results hold also for elliptic problems without any monotonicity condition for the nonlinearity g, since the Leray-Schauder or coincidence degree can now be used.

4.3 Existence of multiple solutions

Assume again that $g = g(s)$ and the assumptions of the preceding section hold but in this case there is only one crossed eigenvalue λ_q with multiplicity $m(\lambda_q) = 1$. Hence the crossing space H_2 is one dimensional. We are now able to improve the result of Theorem 4.1. Indeed, we have

Theorem 4.2. *If $\dim H_2 = 1$, then the equation* (SWE) *admits at least two nontrivial weak solutions u_1 and u_2 in $D(L)$ with the property $(P_2 u_1, P_2 u_2) < 0$.*

We already know by Theorem 4.1 that the equation (SWE) have at least one nontrivial weak solution. Denote by ψ_q the only eigenvector of L corresponding to the crossed eigenvalue λ_q. Hence the crossing space $H_2 = \mathrm{sp}\{\psi_q\}$. The idea of the proof is to show that there exist two open bounded sets G_+ and G_- such that $\overline{G}_+ \cap \overline{G}_- = \emptyset$, $0 \notin \overline{G}_+ \cup \overline{G}_-$ and

$$\deg(L - N, G_\pm, 0) \neq 0.$$

We shall consider homotopy of the form

$$F_\mu := L - \mu N - (1-\mu)(\lambda_q I - P_2), 0 \leq \mu \leq 1,$$

where P_2 is the orthogonal projection to H_2. Note that the only solution for equation

$$Lu - \lambda_q u + P_2 u = \pm\psi_q$$

is $u = \pm\psi_q$. We shall construct the sets G_+ and G_- in such a way that $\psi_q \in G_+$, $-\psi_q \in G_-$ and

$$F_\mu(u) \neq (1-\mu)(\pm\psi_q) \quad \text{for all } 0 \leq \mu \leq 1 \text{ and } u \in \partial G_\pm \cap D(L).$$

We need a priori bounds for the solution set

$$S = \{u \in D(L) | F_\mu(u) = (1-\mu)w \text{ for some } 0 \leq \mu < 1\},$$

where $w = \pm\psi_q$. Note that $0 \notin S$ but $w = \pm\psi_q \in S$. Crucial to the estimates is the behavior of the P_2-component. To this end we consider separately the cases $(P_2 u - w, P_2 u) \geq 0$ and $(P_2 u - w, P_2) < 0$. Denote

$$S_p = \{u \in S | (P_2 u - w, P_2 u) \geq 0\} = \{u \in S | P_2 u = rw, r \leq 0 \text{ or } r \geq 1\}$$

and

$$S_n = \{u \in S | (P_2 u - w, P_2 u) < 0\} = \{u \in S | P_2 u = rw, 0 < r < 1\}.$$

We obtain

Lemma 4.3. *There exist constants $R > 1$ and $0 < \rho < 1$ such that*
(a) $S_p \subset \{u \in H | \|u\| < R\}$ and
(b) $S_n \subset \{u \in H | \|u\| < R, P_2 u = rw, r > \rho\}$.

The proof of Lemma 4.3 is based on similar argumentation as the proof of Lemma 4.2, see [Be-Mu4], [Be-Be2]. It is easy to see that the desired sets are obtained by setting

$$G_\pm = \{u \in H | \|u\| < R, P_2 u = \pm r\psi_q, r > \rho\}.$$

Hence we find solutions $u_1 \in G_+$ and $u_2 \in G_-$ of (SWE) by homotopy argument. Clearly $P_2 u_1 = r_1\psi_q$ and $P_2 u_2 = -r_2\psi_q$ for some $r_1, r_2 > \rho$. Thus the second assertion of Theorem 4.2 is valid, i.e., $(P_2 u_1, P_2 u_2) < 0$. Notice that in case g is odd in s it is possible that $u_2 = -u_1$ and we obtain only one pair $\{u, -u\}$ of nontrivial solutions.

The result of Theorem 4.2 seems to be quite restricted. However, in many cases it can be used to obtain multiple solutions by reducing the original problem to some suitable invariant subspace. We shall close this section by an example which will illuminate this method. We shall use the fact that no assumption of monotonicity on the nonlinearity g is needed in the proof of Lemma 4.3.

Example 4.1. Assume that $g = g(s)$ and it interacts with the eigenvalues 11, 12, 13, 15, 16, 17, 19, 20 and 21 of L, that is

$$9 < \alpha \leq l_\infty < 11 < \cdots < 21 < l_0 \leq \beta < 22.$$

Counting the multiplicities we find out that the dimension of the crossing space H_2 is 23. In the spirit of the work by Coron [Co] (see also [Ve-Sh]) we define the subspace V by setting

$$\begin{aligned} V &= \{u \in H | u(t+\pi, \pi - x) = u(t,x) \text{ for a.a. } t \in]0,\pi[, x \in]0,\pi[\} \\ &= \overline{\text{sp}}\{\psi_{jk} : j \in \mathbb{Z}_+, k \in \mathbb{Z} \text{ with } j+k \text{ is odd }\}. \end{aligned}$$

Now the wave operator L is *completely reduced* by V, i.e.,

$$P_V(D(L)) \subset D(L) \quad \text{and} \quad LP_V u = P_V L u$$

for all $u \in D(L)$, where $P_V : H \to V$ is the orthogonal projection. The restrictions L_V and $L_{V^\perp}$ of L into the spaces V and $V^\perp$, respectively, inherit the properties of L and $\sigma(L) = \sigma(L_V) \cup \sigma(L_{V^\perp})$. The reduced spectrum is now

$$\sigma(L_V) = \{j^2 - k^2 : j \in \mathbb{Z}_+, k \in \mathbb{Z}, j+k \text{ odd}\}.$$

Interesting fact here is that $0 \notin \sigma(L_V)$, and hence L_V is invertible with compact inverse L_V^{-1}. This makes it possible to apply the Leray-Schauder theory, where the nonlinearity need not to be monotone. Since $N(V) \subset V$, any solution of the reduced equation

$$L_V u - N_V(u) = 0, \quad u \in D(L) \cap V,$$

is also weak solution for the original equation (SWE). It is not hard to see that $\dim(V \cap H_2) = 16$. Denote

$$\begin{aligned} \Lambda &= \{(j,k) \in \mathbb{Z}_+ \times \mathbb{Z} : \alpha < j^2 - k^2 < \beta, j+k \text{ is odd}\} \\ &= \{(6,\pm 5), (7,\pm 6), (8,\pm 7), (4,\pm 1), (9,\pm 8), (10,\pm 9), (11,\pm 10), (5,\pm 2)\}. \end{aligned}$$

For each $k \in \{6,7,8,9,10\}$ we associate a further subspace

$$V_k = \{u \in V : u(t+\frac{2\pi}{k},x) = u(t,x) \text{ for a.a. } t \in]0,2\pi - \frac{2\pi}{k}[, \ x \in]0,\pi[$$
$$\text{and } u(2\pi - t, x) = u(t,x) \text{ for all } (t,x) \in \Omega\}$$
$$= \overline{\text{sp}}\{\psi_{jm} : j \in \mathbb{Z}_+, \frac{j}{-m} \in \mathbb{Z}_+, j+m \text{ is odd}\}$$

Reasoning as before we see that the solutions of the reduced equations $L_{V_k}u - N_{V_k}u = 0, u \in D(L) \cap V_k, k = 6,7,8,9,10$, are also solutions for the original equation. Clearly $\sigma(L_{V_6}) \cap]\alpha, \beta[$ contains only one eigenvalue $7^2 - (-6)^2 = 13$ with multiplicity one. Thus by Theorem 4.2 we find two nontrivial solutions for (SWE). The same conclusion hold for $k = 7, 8, 9$ and 10, respectively. Some of these solutions may be caused by a shift $t \to t + \tau$ (they are physically equivalent). Excluding those we get the final result: the equation (SWE) admit at least five different nontrivial solutions even without any monotonicity condition for the nonlinearity g. If we assume that g is odd, then it is possible to find altogether eight pairs of nontrivial solutions.

4.4 Existence of multiple solutions–resonance case

In this section we consider existence of multiple solutions for the equation

$$\text{(RE)} \qquad \begin{cases} u_{tt} - u_{xx} - \lambda u - g(t,x,u) = h, & (t,x) \in \mathbb{R} \times (0,\pi), \\ u(t,0) = u(t,\pi) = 0, & t \in \mathbb{R}, \\ u(\cdot,x) \text{ is } 2\pi\text{-periodic for all } x \in (0,\pi), \end{cases}$$

where $\lambda \in \sigma(L)$ is fixed and positive. We assume that the Caratheodory function g is such that the corresponding Nemytski operator N has bounded range and the map $\lambda I + N$ is of class (S_+). Naturally $h \in H = L^2(\Omega)$ and we can write $h = h_1 + h_2$, where $h_1 \in \text{Ker}(L - \lambda I)$ and $h_2 \in \text{Ker}(L - \lambda I)^\perp$. Denote by P_λ the orthogonal projection from H to $\text{Ker}(L - \lambda I)$.

We shall study for which $h \in H$ the resonance equation (RE) has weak solutions. We assume that there exists a function $J_N : H \to \mathbb{R}_+$ such that

$$\text{(A1)} \qquad \liminf((N(u_k), P_\lambda u_k) - J_N(P_\lambda u_k)) > 0,$$

$$\text{(A2)} \qquad J_N(w) \geq \limsup(N(u_k), w),$$

for all $w \in H$ and all sequences $\{u_k\} \subset H$ with $\|P_\lambda u_k\| \to \infty$ and $\|(I - P_\lambda)u_k\| \leq$ constant. In applications J_N will be the recession function introduced by Brezis and Nirenberg in [Br-Ni]. Using J_N we define the set

$$\Gamma = \{h_1 \in \text{Ker}(L - \lambda I) | J_N(w) \geq -(h_1, w) \text{ for all } w \in \text{Ker}(L - \lambda I)\}.$$

Obviously $0 \in \Gamma$ and Γ is a closed bounded subset of $\text{Ker}(L - \lambda I)$. Our main result is

Theorem 4.3. *For each fixed $h_2 \in \operatorname{Ker}(L - \lambda I)^\perp$ there exists an open set $\Gamma_{h_2} \subset \operatorname{Ker}(L - \lambda I)$ containing Γ such that*

(a) equation (RE) *has at least one weak solution for all $h = h_1 + h_2$ with $h_1 \in \Gamma_{h_2}$,*

(b) equation (RE) *has at least two different solutions for all $h = h_1 + h_2$ with $h_1 \in \Gamma_{h_2} \backslash \Gamma$.*

The proof of assertion (a) is based on the homotopy

$$F_\mu(u) := Lu - \lambda u - \mu N(u) - (1 - \mu) P_\lambda u \neq \mu h, \quad 0 \leq \mu \leq 1, u \in \partial G,$$

where G is a "box" of the form $G_n = \{u \in H | \|P_\lambda u\| < n, \|(I - P_\lambda)u\| < M + 1\}$ with n a positive integer and $M > 0$ a constant. Let $G = G_{n_0}$. In view of Lemma 4.1 we get

$$\deg(L - \lambda I - N, G_{n_0}, h) = \deg(L - \lambda I - P_\lambda, G_{n_0}, 0) \neq 0$$

implying that (a) holds. To prove the second assertion we shall show that there exists $n_1 > n_0$ such that $\deg(L - \lambda I - N, G_{n_1}, h) = 0$. Thus by the additivity property of degree

$$\deg(L - \lambda I - N, G_{n_1} \backslash \overline{G}_{n_0}, h) = - \deg(L - \lambda I - N, G_{n_0}, h) \neq 0$$

giving assertion (b).

We shall sketch the details of the proof of Theorem 4.3. Let $h_2 \in \operatorname{Ker}(L - \lambda I)^\perp$ be fixed. Since the range of N is bounded it is easy to see that for any solution u of $F_\mu(u) = \mu(h_1 + h_2)$ there exists a constant $M > 0$ depending only on h_2 such that

$$\|(I - P_\lambda)u\| \leq M.$$

We show first that there exists n_0 such that $F_\mu(u) \neq \mu(h_1 + h_2)$ for all $u \in \partial G_n, \mu \in [0,1], h_1 \in \Gamma$ and $n \geq n_0$. Indeed, assuming the contrary we have $F_{\mu_k}(u_k) = \mu_k(h_1^{(k)} + h_2)$ with $u_k \in \partial G_{n_k}, \mu_k \in [0,1], h_1^{(k)} \in \Gamma$ and $n_k \to \infty$. Clearly $\|P_\lambda u_k\| = n_k$ and we can use (A1) together with the definition of Γ to obtain a contradiction. Hence $\deg(L - \lambda I - N, G_{n_0}, h_1 + h_2) \neq 0$ and there exists a weak solution $u \in G_{n_0}$ for (RE) for all $h_1 \in \Gamma$. Moreover, every $h_1 \in \Gamma$ has an open neighbourhood $V(h_1) \subset \operatorname{Ker}(L - \lambda I)$ such that $\deg(F_1, G_{n_0}, w + h_2) \neq 0$ for all $w \in V(h_1)$. The open set Γ_{h_2} in $\operatorname{Ker}(L - \lambda I)$ is now obtained by taking $\Gamma_{h_2} = \cup_{h_1 \in \Gamma} V(h_1)$.

Choose now $h_1 \in \Gamma_{h_2} \backslash \Gamma$. Since $h_1 \notin \operatorname{Im}(L - \lambda I)$ and the range of N is bounded we can find $r > 0$ such that

$$Lu - \lambda u - N(u) \neq (1 + r)h_1 + h_2 \quad \text{for all } u \in D(L).$$

Hence $\deg(F_1, G_n, (1+r)h_1 + h_2) = 0$ for all n and reasoning as above it is easy to prove that there exists $n_1 > n_0$ such that $F_1(u) \neq (1+\theta r)h_1 + h_2$ for all $u \in \partial G_{n_1}$ and $\theta \in [0,1]$. Thus

$$\deg(F_1, G_{n_1}, h_1 + h_2) = \deg(F_1, G_{n_1}, (1+r)h_1 + h_2) = 0$$

and in view of the preceding remarks the proof of assertion (b) is now completed.

The result of Theorem 4.3 is originally proved by Hofer in [Ho] using Galerkin approximations and Brouwer degree. He also gives a concrete example where the requirements of Theorem 4.3 are met.

4.5 Bifurcation at infinity

In this section we shall deal with the existence of solutions (u, λ) of equation

$$\text{(BE)} \qquad \begin{cases} u_{tt} - u_{xx} - \lambda u - g(\lambda, t, x, u) = 0, (t,x) \in \mathbb{R} \times (0, \pi), \\ u(t,0) = u(t,\pi) = 0 \quad \text{for all } t \in \mathbb{R}, \\ u(\cdot, x) \quad \text{is } 2\pi\text{-periodic for all } x \in (0,\pi), \end{cases}$$

where $\lambda \in \mathbb{R}_+$ and the nonliner part g is sublinear with respect to u. We are interested in solutions (u, λ) where u is "large" which leads us to the study of asymptotic bifurcation points of (BE).

We shall assume that the function $(\lambda, t, x, s) \to g(\lambda, t, x, s)$ from $\mathbb{R}_+ \times \mathbb{R} \times [0,\pi] \times \mathbb{R}$ to $\mathbb{R}$ is continuous in (λ, s) for a.a. (t,x) and measurable in (t,x) for all λ and s, 2π-periodic in t and that for each bounded interval $[a,b] \subset \mathbb{R}_+$ there exist constants $\beta > 0, 0 < \alpha < 1$ and a function $h_0 \in L^2(\Omega)$ such that

$$|g(\lambda, t, x, s)| \leq \beta |s|^\alpha + h_0(t,x) \quad \text{for a.a. } (t,x) \in \Omega \text{ and } (\lambda, s) \in [a,b] \times \mathbb{R}.$$

Then the corresponding Nemytski operator $N(\lambda, \cdot)$ is sublinear in the sense that

$$\frac{\|N(\lambda, u)\|}{\|u\|} \to 0 \quad \text{as } \|u\| \to \infty \text{ uniformly in } \lambda \text{ on } [a,b].$$

Moreover we assume that there exists a constant $\theta > 0$ such that

$$\{g(\lambda, t, x, s_1) - g(\lambda, t, x, s_2)\}(s_1 - s_2) \geq (\theta - \lambda)|s_1 - s_2|^2$$

for all $s_1, s_2 \in \mathbb{R}, \lambda \in [a,b]$ and a.a. $(t,x) \in \Omega$. It is easy to see that the map $(\lambda, u) \to \lambda u + N(\lambda, u)$ from $[a,b] \times H$ to H defines a bounded homotopy of class (S_+).

We recall that λ is said to be *an asymptotic point of bifurcation* for equation (BE), if there exists a sequence of solutions $\{(\mu_k, u_k)\}$ such that $\mu_k \to \lambda$ and $\|u_k\| \to \infty$. A standard necessary condition for λ to be an asymptotic point of bifurcation is that λ belongs to the spectrum of the wave operator L, see [Ber3]. Hovever, it is not generally a sufficient condition, see [De]. We have the following

Theorem 4.4. *Assume that $\lambda_q \in \sigma(L) \cap (a,b)$ and the multiplicity $m(\lambda_q)$ is odd. Then λ_q is an asymptotic point of bifurcation for equation* (BE).

Proof. Let r_1 and r_2 be two fixed numbers such that $a \le r_1 < \lambda_q < r_2 \le b$ and $[r_1, r_2] \cap \sigma(L) = \{\lambda_q\}$. We shall show that there exists $R > 0$ such that

$$Lu - r_i u - sN(r_i, u) \neq 0$$

for all $u \in D(L), \|u\| \ge R, s \in [0,1]$ and $i = 1, 2$. Indeed, assume the contrary. Then we can find sequences $\{u_k\} \subset D(L)$ and $\{s_k\} \subset [0,1]$ such that $\|u_k\| \to \infty$ and

$$Lu_k - r_i u_k - s_k N(r_i, u_k) = 0.$$

Denote $v_k = \frac{u_k}{\|u_k\|}$. At least for subsequence we can write $v_k \rightharpoonup v$. Hence

$$Lv_k - r_i v_k = s_k \frac{N(r_i, u_k)}{\|u_k\|} \to 0$$

implying by the properties of L that $v_k \to v \in D(L)$ and $Lv - r_i v = 0$. Since $r_i \notin \sigma(L)$ we have $v = 0$, which is a contradiction. Thus we can use homotopy argument together with Lemma 4.1 to obtain

$$\begin{aligned}\deg(L - r_i I - N(r_i, \cdot), B_R(0), 0) &= \deg(L - r_i I, B_R(0), 0)\\ &= (-1)^{\sum\limits_{0<\lambda_j<r_i} m(\lambda_j)}\end{aligned}$$

By the fact that $m(\lambda_q)$ is odd we get

$$\deg(L - r_1 I - N(r_1, \cdot), B_R(0), 0) \neq \deg(L - r_2 I - N(r_2, \cdot), B_R(0), 0).$$

Now we can use the homotopy argument in reverse order. Indeed, since the degrees above are not equal, there necessarily exist $\mu \in (r_1, r_2)$ and $\bar{u} \in \partial B_R(0)$ such that

$$L\bar{u} - \mu\bar{u} - N(\mu, \bar{u}) = 0.$$

Since r_1 and r_2 can be chosen arbitrarily close to λ_q and $R > 0$ taken arbitrarily large we conclude the existence of a sequence $\{(\mu_k, u_k)\}$ of solutions such that $\mu_k \to \lambda_q$ and $\|u_k\| \to \infty$. Hence the proof is complete. □

Note that the eigenvalues of the wave operator are of the form $j^2 - k^2$, where $j \in \mathbb{Z}_+$ and $k \in \mathbb{Z}$. Thus the multiplicity of an eigenvalue λ is odd if and only if $\lambda = j^2$ for some $j \in \mathbb{Z}_+$.

Theorem 4.4 is an example of local bifurcation result. Corresponding local result, where the sublinearity of N is replaced by condition

$$\frac{\|N(\lambda, u)\|}{\|u\|} \to 0 \quad \text{as } u \to 0 \text{ uniformly in } \lambda \text{ in } [a,b]$$

can be found in [Ber3]. The above condition, however, is not relevant for a Nemytski operator in L^2-setting. It is possible to study the the global behavior of the branches of bifurcated solutions using degree argument. A result in this direction are given in [Ber3]. In some special cases the asymptotic bifurcation may occur near an eigevalue with even multiplicity or there may be several different branches of solutions emanating from the same eigenvalue, see [Be-Mu9].

5. Generalization of the Jordan-Brouwer separation theorem

We shall close this paper by considering briefly the generalization of the famous Jordan-Brouwer separation theorem. In its simplest form it is known as the Jordan curve theorem and it states that any closed curve γ in $\mathbb{R}^2$ homeomorphic to unit circle divides the plane into two open connected components D_1 and D_2 such that $\gamma = \partial D_1 = \partial D_2$. More generally we have

Jordan-Brouwer theorem. *Let K_1 and K_2 be closed bounded subsets in $\mathbb{R}^n$, $n \geq 2$, and $h : K_1 \to K_2$ a homeomorphism. Then the sets $\mathbb{R}^n \backslash K_1$ and $\mathbb{R}^n \backslash K_2$ have the same number of open connected components or both have countably infinitely many.*

A direct analytical proof of the above JB-theorem is quite technical. Another way is to use algebraic topology. We are interested in a variant of analytical proof which is based on the use on topological degree, cf. [Sch], [Llo]. Denote by $\mathcal{C}$ the class of continuous mappings in the space $\mathbb{R}^n$. Thus $h, h^{-1} \in \mathcal{C}$. A careful reading of the proof of JB-theorem in $\mathbb{R}^n$ reveals that it is based on the following facts:

(a) The functions h and h^{-1} have continuous extensions $\tilde{h}$ and $\tilde{h}^{-1}$, respectively, to the whole of $\mathbb{R}^n$.

(b) There exists a classical topological degree theory for the class $\mathcal{C}$.

(c) The composite of two continuous functions is also continuous.

(d) The Brouwer degree for a composite function can be calculated by using so called "product formula".

We recall the exact meaning of the product formula or multiplication theorem in $\mathbb{R}^n$. Let $G \subset \mathbb{R}^n$ be an open bounded set and $f : \overline{G} \to \mathbb{R}^n$ a continuous map. Let $\Delta_j (j = 1, 2, \dots)$ be the open bounded components of $\mathbb{R}^n \backslash f(\partial G)$ and Δ_∞ the unique unbounded component. Denote $D = \cup_{j \neq \infty} \Delta_j$. If $g : \mathbb{R}^n \to \mathbb{R}^n$ is a continuous map and $y \notin g \circ f(\partial G)$, then

$$\text{(PF)} \qquad d_B(g \circ f, G, y) = \sum_{j \neq \infty} d_B(g, \Delta_j, y) d_B(f, G, \Delta_j).$$

There are at most countably infinitely many components of $\mathbb{R}^n \backslash f(\partial G)$, since $\mathbb{R}^n$ is separable. Note that the summation in (PF) is finite since $g^{-1}(y) \cap D$ is a compact set, so that, being covered by the disjoint open sets Δ_j, it meets only finite number of them and the other terms in (PF) vanish. The product formula seems not to be an automatic consequence of the other properties of topological degree. It is valid for Leray-Schauder type maps in infinite dimensional real Banach space X. It is easy to to see that for a Leray-Schauder type injection $F = I - C$, also its inverse is of the type $F^{-1} = I - \hat{C}$, where $\hat{C}$ is compact. Similarly the composite function of two Leray-Schauder maps is in the same class. For the proof of product formula for Brouwer and Leray-Schauder degrees we refer to [Llo] or [Sch].

The condition (a) is valid by Tietzes extension theorem. Analogous property for a Leray-Schauder type map follows from the result due to Dugundji:

Any compact map C from a subset of a real Banach space X into itself admits a compact extension $\tilde{C} : X \to X$.

Hence in view of (a)-(d) we obtain

JB-theorem in Leray-Schauder case. *Let K_1 and K_2 be closed bounded subsets of a real Banach space X and $h = I - C : K_1 \to K_2$, where C is compact, a homeomorphism. Then the sets $X \backslash K_1$ and $X \backslash K_2$ have the same number of open connected components or both have countably infinitely many.*

There are some special cases where the product formula hold. For instance, in [Ber1] it is proved that if $F = I - C : \overline{G} \to X$ is a Leray-Schauder type map, where G is an open bounded set in a real reflexive Banach space X, and $T : X \to X^*$ bounded and of class (S_+), then $T \circ F$ is of class (S_+) and

$$d(T \circ F, G, y) = \sum_{j \neq \infty} d(T, \Delta_j, y) d_{LS}(F, G, \Delta_j)$$

for any $y \in X^* \backslash (T \circ F)(\partial G)$. For other generalizations and extensions of the product formula, see [Mo-Ha], [Wo].

Let us now study more closely the possibility to generalize the JB-theorem for mappings of class (S_+). Let H be a real separable Hilbert space and K_1, K_2 closed bounded subsets of H. Assume that $h : K_1 \to K_2$ is a homeomorphism. In general, the continuity is not sufficient condition to guarantee that JB-theorem is valid. For example, if K_1 is the unit sphere in H and K_2 the unit sphere in some closed infinite dimensional subspace of H, then they are homeomorphic but JB-theorem does not hold. Assume that the homeomorphism h is of class (S_+). Then it is easy to see that also $h^{-1} : K_2 \to K_1$ is of class (S_+). However, we encounter several difficulties. First of all, it is not known whether or not h and h^{-1} have (S_+)-type extensions to the whole of H. Secondly, the composite map of two (S_+)-maps is not necessarily in the same class; for counterexample, see [Ber2]. This is, of course, connected to the validity of product formula (PF) – the lefthand side may not be well-defined. Part of these problems are solved in [Ber2], where a new degree theory is constructed for mappings $S : H \to H$, which have a representation $S = T \circ F$, where T, F are bounded continuous and of class (S_+). The representation is not unique but the value of the degree is independent of it. Note that the new class contains all (S_+)-maps, since we can always write $F = I \circ F$ and the identity is of class (S_+). It is proved in [Ber2] that the product formula hold, that is, if $F : \overline{G} \to H$ and $T : H \to H$ are bounded continuous (S_+)-maps, then with obvious notations

$$d(T \circ F, G, y) = \sum_{j \neq \infty} d(T, \Delta_j, y) d(F, G, \Delta_j)$$

for all $y \notin (T \circ F)(\partial G)$. Hence all requirements except (a), existence of admissible extensions for h and h^{-1}, are satisfied. However, we get the following partial result.

Theorem 5.1. *Let K_1 and K_2 be closed bounded subsets of the Hilbert space H. Assume that there exists a homeomorphism $h : K_1 \to K_2$ of the form*

$$h = I - T - C,$$

where $C : K_1 \to H$ is compact and $T : K_1 \to H$ satisfies a contraction condition

$$\|T(u) - T(v)\| \leq k\|u - v\| \quad \text{for all } u, v \in K_1$$

with some constant $0 < k < 1$. Then $H\backslash K_1$ and $H\backslash K_2$ have the same number of open components or both have countably infinitely many.

Proof. It is easy to see that $h = I-T-C$ is of class (S_+). By the preceding discussion it is sufficient to prove that h and h^{-1} have admissible, i.e., (S_+)-type, extensions to all of H. Indeed, the map T can be extended to all of H in such a way that the extension $\tilde{T}$ satisfies contraction condition

$$\|\tilde{T}(u) - \tilde{T}(v)\| \leq k\|u - v\| \quad \text{for all } u, v \in H$$

with the same constant $k \in (0,1)$ as T [Sch p.21] . We also know that the map C admits a compact extension $\tilde{C}$ to all of H. Hence the mapping $\tilde{h} = I-\tilde{T}-\tilde{C} : H \to H$ is the required extension of h. In order to find the extension of h^{-1} we notice that $I-T$ is strongly monotone and hence an injection of class (S_+). Thus also $(I-T)^{-1}$ belongs to class (S_+). It is easy to see that

$$h^{-1}(w) = (I-T)^{-1}(w + C \circ h^{-1}(w)) \quad \text{for all } w \in K_2.$$

Since the map $C \circ h^{-1} : K_2 \to H$ is compact, it has a compact extension $C_0 : H \to H$. Hence we can define

$$\tilde{h}^{-1} = (I - \tilde{T})^{-1}(I + C_0) : H \to H,$$

which is an (S_+)-type extension of h^{-1}. Thus the proof is complete. □

REFERENCES

[Ama1] H.AMANN, *Saddle points and multiple solutions of differential equations*, Math.Z. **169** (1979), 127–166.

[Ama2] H.AMANN, *On the unique solvability of semilinear operator equations in Hilbert spaces*, J.Math.Pures Appl. **61** (1982), 149–175.

[Am-We] H.AMANN AND S.WEISS, *On the unqueness of the topological degree*, Math.Z. **130** (1973), 39–54.

[Am-Ze] H.AMANN AND E.ZEHNDER, *Multiple periodic solutions for a class of nonlinear autonomous wave equations*, Houston J. Math. **7** (1981), 147–174.

[Be-Be1] A.K.BEN-NAOUM AND J.BERKOVITS, *On the existence of periodic solutions for semilinear wave equation on ball in* $\mathbb{R}^n$ *with the space dimension* n *odd*, Nonlinear Analysis, TMA **24** (1994), 241–250.

[Be-Be2] A.K.BEN-NAOUM AND J.BERKOVITS, *Nontrivial solutions for some semilinear problems and applications to wave equations on ball and on sphere*, TMNA **3** (1994), No 2.

[Ber1] J.BERKOVITS, *On the degree theory for mappings of monotone type*, Ann.Acad.Sci.Fenn.Ser.A1, Dissertationes **58** (1986).

[Ber2] J.BERKOVITS, *Topological degree theory and multiplication theorem for a class of nonlinear mappings*, Bull.London Math.Soc. **23** (1991), 596–606.

[Ber3] J.BERKOVITS, *Some bifurcation results for a class of semilinear equations via topological degree method*, Bull. Soc. Math. Belg. **44** (1992), 237–247.

[Be-Mu1] J.BERKOVITS AND V.MUSTONEN, *On the topological degree for mappings of monotone type*, Nonlinear Analysis, TMA **10** (1986), 1373–1383.

[Be-Mu2] J.BERKOVITS AND V.MUSTONEN, *Nonlinear mappings of monotone type, Part I: Classification and degree theory*, Report **2** (1988), Mathematics, Univ. of Oulu, 53p.

[Be-Mu3] J.BERKOVITS AND V.MUSTONEN, *An extension of Leray-Schauder degree and applications to nonlinear wave equations*, Differential Integral Equations **3** (1990), 945–963.

[Be-Mu4] J.BERKOVITS AND V.MUSTONEN, *On multiple solutions for a class of semilinear wave equations*, Nonlinear Analysis, TMA **16** (1991), 421–434.

[Be-Mu5] J.BERKOVITS AND V.MUSTONEN, *On the existence of multiple solutions for semilinear equations with monotone nonlinearities crossing finite number of eigenvalues*, Nonlinear Analysis, TMA **17** (1991), 399–412.

[Be-Mu6] J.BERKOVITS AND V.MUSTONEN, *An application of the topological degree theory for semilinear equations with strongly monotone nonlinearities*, Bull. London Math. Soc **23** (1991), 470–476.

[Be-Mu7] J.BERKOVITS AND V.MUSTONEN, *Topological degree for perturbations of linear maximal monotone mappings and applications to a class of parabolic problems*, Rendiconti di Matematica, Serie VII, Vol. 12 (Roma 1992), 597–621.

[Be-Mu8] J.BERKOVITS AND V.MUSTONEN, *Existence and multiplicity results for nonlinear beam equations*, Collequia Mathematica Societatis Janos Bolyai, Vol. 62, Differential Equations, North-Holland, Budapest, (1991) 1994, 49–63.

[Be-Mu9] J.BERKOVITS AND V.MUSTONEN, *On the asymptotic bifurcation for a class of wave equations with sublinear nonlinearity*, Partial differential equations (J.Wiener and J.K.Hale ed.), Pitman Research Notes in Mathematics, Series 273 (1992), 159–162.

[Be-Mu10] J.BERKOVITS AND V.MUSTONEN, *On the resonance for semilinear equations with normal linear part*, J.Math.Maroc **2** (1994), 115–123.

[Be-Mu11] J.BERKOVITS AND V.MUSTONEN, *Monotonicity methods for nonlinear evolution equations*, Nonlinear Analysis, TMA 27 (1996), 1397–1405.

[Be-Ti] J.BERKOVITS AND M.TIENARI, *Topological degree for some classes of multis with applications to hyperbolic and elliptic problems involving dicontinuous nonlinearities*, Dynamic Systems and Applications **5** (1996), 1–18.

[Br-Ni] H.BREZIS AND L.NIRENBERG, *Characterizations of the ranges of some nonlinear operators and applications to boundary value problems*, Ann. Scuola Norm. Pisa **(4)5** (1978), 225–326.

[Brow] F.E.BROWDER, *Fixed point theory and nonlinear problems*, Proc. Sympos. Pure Math. **39** (1983), 49–87.

[Br-To] F.E.BROWDER AND B.A.TON, *Nonlinear functional equations in Banach spaces and elliptic super-regularization*, Math.Z **105** (1968), 177–195.

[Brou] L.E.J.BROUWER, *Uber Abbildungen von Mannichfaltigkeiten*, Math. Ann. **71** (1912), 97–115.

[Cor] J.M.CORON, *Periodic solutions of a nonlinear wave equation without assumption of monotonicity*, Math. Ann. **262** (1983), 273–285.

[Cro] J.CRONIN, *Fixed points and topological degree in nonlinear analysis*, vol. 11, American Mathematical Society, Providence, Rhode Island, 1964.

[De] K.DEIMLING, *Nonlinear functional analysis*, Springer-Verlag, Berlin, 1985.

[Dr-Hu] P.DRABEK AND Y.X.HUANG, *Bifurcation problems for the p-Laplacian in* $\mathbb{R}^n$, Trans.Amer.Math.Soc. (to appear).

[Dr] P.DRABEK, *On the bifurcation for a class of degenerate equations*, Ann.Mat.Pura Appl. **159** (1991), 1–16.

[Fu] L.FUHRER, *Ein elementarer analytischer Beweis zur Eindeutigkeit des Abbildungsradet im* $\mathbb{R}^n$, Math.Nachr. **54** (1972), 259–267.

[Hi] N.HIRANO, *On the existence of multiple solutions for a nonlinear wave equation*, J. Differential Equations **71** (1988), 334–347.

[Ho] H.HOFER, *A multiplicity result for a class of nonlinear problems with applications to a nonlinear wave equation*, J.Nonlinear Analysis **5** (1981), 1–11.

[Kit] A.KITTILA, *On the topological degree for a class of mappings of monotone type and applications to strongly nonlinear elliptic problems*, Ann.Acad.Sci.Fenn.Ser.A1, Dissertationes **91** (1994).

[La-Mc] A.C.LAZER AND P.J.McKENNA, *Critical point theory and boundary value problems with nonlinearities crossing multiple eigenvalues*, Comm. Partial Diff. Equations **11** (1986), 1653–1676.

[LS] J.LERAY AND J.SCHAUDER, *Topologie et equationes functionelles*, Ann. Sci. Ecole Norm. Sup. **51** (1934), 45–78.

[Llo] N.G.LLOYD, *Degree Theory*, Cambridge University Press, Cambridge, 1978.

[Ma1] J.MAWHIN, *Equivalence theorems for nonlinear operator equations and coincidence degree theory for some mappings in locally convex topological spaces*, J.Differential Equations **12** (1972), 610–636.

[Ma2] J.MAWHIN, *Topological degree methods in nonlinear boundary value problems*, CBMS Regional Conf. Ser. in Math.,no. 40, Amer. Math. Soc., Providence, R.I., 1979.

[Mo-Ha] H.MONCH AND G.F.VON HARTEN, *The product formula for the topological degree of strict γ-contractions*, Manuscripta Math **23** (1978), 113–123.

[Ra] P.H.RABINOWITZ, *Some global results for nonlinear eigenvalue problems*, J.Funct.Anal. **7** (1971), 487–513.

[Sch] J.T.SCHWARTZ, *Nonlinear Functional Analysis*, Gordon and Breach, New York, 1969.

[Tie] M.TIENARI, *A degree theory for a class of mappings of monotone type in Orlicz-Sobolev spaces*, Ann.Acad.Sci.Fenn.Ser.A1, Dissertationes **97** (1994).

[Wo] S.F.WONG, *A product formula for the degree of A-proper maps*, J.Funct.Anal. **10** (1972), 361–371.

Juha Berkovits
Department of Mathematical Sciences
University of Oulu
FIN-90570 Oulu
Finland

J-P GOSSEZ

Some remarks on the antimaximum principle

1 Introduction

A function u on an interval $[a, b]$ which satisfies $-u'' \geq 0$ on $[a, b]$ achieves its minimum at a or b (because u is concave). If in addition $u(a) = u(b) = 0$, then $u \geq 0$ on $[a, b]$, and in fact either $u \equiv 0$ on $[a, b]$, or $u > 0$ on $]a, b[$ with $u'(a) > 0$ and $u'(b) < 0$. These observations from calculus represent the most elementary form of the *maximum principle*, which is a fundamental result in the theory of elliptic partial differential equations.

Let us consider the Dirichlet problem

$$\begin{cases} -\Delta u = f(x) \text{ in } \Omega, \\ u = 0 \text{ on } \partial\Omega, \end{cases} \tag{1.1}$$

where Ω is a smooth bounded domain in $\mathbb{R}^N$ and Δ is the Laplacian operator. The *maximum principle* for (1.1) states that if $f \in L^\infty(\Omega)$ is ≥ 0 a.e. in Ω,then the solution u of (1.1) (which belongs to $C^1(\bar{\Omega})$) is ≥ 0 in Ω; moreover either $u \equiv 0$ in Ω, or u is > 0 in Ω with $\partial u/\partial n < 0$ on $\partial\Omega$ ($\partial/\partial n$ represents the exterior normal derivative). More generally the same conclusion holds for the problem

$$\begin{cases} -\Delta u = \lambda u + f(x) \text{ in } \Omega, \\ u = 0 \text{ on } \partial\Omega \end{cases} \tag{1.2}$$

provided the real parameter λ satisfies $\lambda < \lambda_1$, where $\lambda_1 > 0$ is the first eigenvalue of the problem

$$\begin{cases} -\Delta u = \lambda u \text{ in } \Omega, \\ u = 0 \text{ on } \partial\Omega. \end{cases} \tag{1.3}$$

In other words, when going from f to the solution u, the sign is preserved and even reinforced : $f \geq 0, \not\equiv 0$ implies $u > 0$. Standard references for the various versions of the maximum principle include [Pr-We], [Gi-Tr].

In 1979, P.Clément and L.Peletier studied problem (1.2) in the case where the parameter λ satisfies $\lambda > \lambda_1$.They derive for certain values of λ a conclusion which is totally opposed to the preceding one : $f \geq 0, \not\equiv 0$ implies $u < 0$, which suggested the name of *antimaximum principle.* In a precise manner : given $f \in L^\infty(\Omega)$ with $f \geq 0$ a.e. in Ω, $f \not\equiv 0$, there exists $\delta > 0$ such that if u is a solution of (1.2) with $\lambda_1 < \lambda < \lambda_1 + \delta$, then $u < 0$ in Ω and $\partial u/\partial n > 0$ on $\partial\Omega$.

The purpose of this Note is to present some recent results dealing with this antimaximum principle. In §2 a relation is exhibited between this principle and the Fučik spectrum of the differential operator. In §3 this principle is extended to the case of the p-Laplacian operator :

$$\Delta_p u = div(|grad\, u|^{p-2} grad\, u), \tag{1.4}$$

where $1 < p < \infty$. Let us also mention a recent work by P.Takáč [Ta] dealing with an abstract version of the antimaximum principle in the setting of ordered Banach spaces.

2 Fučik spectrum

Let us denote by $u^+ = \max(u, 0)$ and $u^- = \max(-u, 0)$ the positive and negative parts of u. The Fučik spectrum is defined as the set Σ of those $(\alpha, \beta) \in \mathbb{R}^2$ such that the problem

$$\begin{cases} -\Delta u = \alpha u^+ - \beta u^- \text{ in } \Omega, \\ u = 0 \text{ on } \partial\Omega \end{cases} \tag{2.1}$$

has a nontrivial solution. The usual spectrum corresponds to the case $\alpha = \beta$.

This generalized spectrum was introduced in the 70's by S.Fučik [Fu] and N.Dancer [Da$_1$] in connection with the study of semilinear problems of the form

$$\begin{cases} -\Delta u = f(x, u) \text{ in } \Omega, \\ u = 0 \text{ on } \partial\Omega. \end{cases} \tag{2.2}$$

The description of this set Σ remains a largely open question. For instance, while it is classical that the usual spectrum (i.e. the intersection of Σ with the diagonal in $\mathbb{R}^2$) is made of a sequence going to infinity, it is not known whether $\Sigma \subset \mathbb{R}^2$ has a nonempty interior (cf. [Da$_2$] for a partial result in that direction).

The beginning of the Fučik spectrum was studied recently in [Cu-Go], [DF-Go]. It is easily seen that Σ is contained in $\{(\alpha, \beta) \in \mathbb{R}^2; \alpha \text{ and } \beta \geq \lambda_1\}$ and contains the straight lines $\lambda_1 \times \mathbb{R}$ and $\mathbb{R} \times \lambda_1$. The existence of a first non trivial curve in Σ going through the second eigenvalue (λ_2, λ_2) and extending to infinity was established in [Cu-Go], [DF-Go]. A variational characterization of this curve in the line of the classical Courant-Hilbert formulas was also given. (Another variant of characterization, using the mountain pass theorem, can also be given, cf. the work of M.Cuesta, D.De Figueiredo, B.Ruf and J.-P. Gossez described in [Cu].) This first nontrivial curve can be written as follows

$$\mathcal{C}_2 = \{(\alpha(t), \beta(t)) \in \mathbb{R}^2; t \in \mathbb{R}\}.$$

It is contained in $\{(\alpha, \beta) \in \mathbb{R}^2; \alpha \text{ and } \beta > \lambda_1\}$, it is symmetric with respect to the diagonal, and the functions $\alpha(t)$, $-\beta(t)$ are continuous, strictly increasing and such that $\alpha(t) \to +\infty$ as $t \to +\infty$ and $\beta(t) \to +\infty$ as $t \to -\infty$.

The asymptotic behaviour of this curve $\mathcal{C}_2$ is connected with the antimaximum principle. In the statement of this principle as given in the introduction, the number $\delta > 0$ depends on f: $\delta = \delta(f)$. Let us say that the **uniform** antimaximum principle holds if there exists $\delta > 0$ such that if u is a solution of (1.2) with $\lambda_1 < \lambda < \lambda_1 + \delta$ and $f \in L^\infty(\Omega)$, $f \geq 0$ a.e., $f \not\equiv 0$, then $u < 0$ in Ω and $\partial u/\partial n > 0$ on $\partial\Omega$. So here δ does not depend on f.

Proposition 2.1. *(cf. [DF-Go])* The uniform antimaximum principle holds if and only if

(2.3) $$\lim_{t \to +\infty} \beta(t) > \lambda_1.$$

Moreover, in this case, the largest number δ admissible in the statement of this uniform principle is equal to

(2.4) $$\lim_{t \to +\infty} \beta(t) - \lambda_1.$$

The situation described in this proposition is in fact rather exceptional. Indeed the strict inequality holds in (2.3) when the dimension $N = 1$ and the Dirichlet boundary condition $u = 0$ is replaced in (1.2), (2.1) by the Neumann boundary condition $\partial u/\partial n = 0$. On the contrary equality holds in (2.3) (and so the uniform version of the antimaximum principle does not hold) in the case of the Dirichlet boundary condition (with N arbitrary), the Neumann boundary condition with $N \geq 2$ (cf. [DF-Go]) and the mixed boundary condition (with N arbitrary) (cf. [Go-Ma]).

3 The p-Laplacian

The p-Laplacian operator (1.4) is a quasilinear generalization of the classical Laplacian (which corresponds to $p = 2$ in (1.4)). It appears e.g. in some questions from fluid mechanics. Its interest comes also from the fact that it is the derivative of the Sobolev norm $\int_\Omega |grad\ u|^p$.

The problem analogous to (1.2) is now written as

(3.1) $$\begin{cases} -\Delta_p u = \lambda |u|^{p-2}u + f(x) \text{ in } \Omega, \\ u = 0 \text{ on } \partial\Omega, \end{cases}$$

and the corresponding eigenvalue problem is

(3.2) $$\begin{cases} -\Delta_p u = \lambda |u|^{p-2}u \text{ in } \Omega, \\ u = 0 \text{ on } \partial\Omega. \end{cases}$$

The study of the spectrum of the p-Laplacian, i.e. the values of $\lambda \in \mathbb{R}$ for which (3.2) has a nontrivial solution, is also a question which remains largely open. Several works in the 80's (in particular those of A.Anane [An]) contributed to show that the first eigenvalue λ_1 of (3.2) has properties which are rather similar to those in the linear case $p = 2$ (simplicity, strictly positive eigenfunction,...). Let us mention that recently A.Anane and N.Tsouli studied the nodal domains of the eigenfunctions and derived a variational characterization of the second eigenvalue λ_2 (cf. [An-Ts]). Let us also observe that, as in the case of the Fučik spectrum, it is not known whether the spectrum of the p-Laplacian has nonempty interior.

A maximum principle was derived for (3.1) when $\lambda < \lambda_1$: if $f \in L^\infty(\Omega)$ is ≥ 0 a.e. and $\not\equiv 0$, then the solution u of (3.1) (which exists, is unique and of class $C^1(\bar{\Omega})$) is > 0 in Ω and satisfies $\partial u/\partial n < 0$ on $\partial\Omega$ (cf. [Va], [Fl-He-dT]). And recently the antimaximum principle was also extended to the p-Laplacian : given $f \in L^\infty(\Omega)$, ≥ 0 a.e., $\not\equiv 0$, there exists $\delta > 0$ such that if u is a solution of (3.1) with $\lambda_1 < \lambda < \lambda_1 + \delta$, then $u < 0$ in Ω and $\partial u/\partial n > 0$ on $\partial\Omega$ (cf. [Fl-Go-Ta-dT$_1$]).

The proof of this antimaximum principle for (3.1) is very different from that of [Cl-Pe] relative to the linear case $p = 2$. It is based on the following nonexistence result of independent interest.

Proposition 3.1. *(cf. [Fl-Go-Ta-dT$_1$])* Let $f \in L^\infty(\Omega)$, ≥ 0 a.e. and $\not\equiv 0$. Then (3.1) with $\lambda = \lambda_1$ has no solution.

Let us recall here the classical result of Fredholm alternative when $p = 2$. Let $f \in L^\infty(\Omega)$; then problem (1.2) with $\lambda = \lambda_1$ has a solution if and only if $\int_\Omega f(x)\varphi_1(x) = 0$, where φ_1 is the eigenfunction (> 0 in Ω) associated with λ_1. In particular (1.2) with $\lambda = \lambda_1$ has no solution if $f \not\equiv 0$ does not change sign. Proposition 3.1 appears in this respect as a first step towards an extension of this Fredholm alternative to the quasilinear case $p \neq 2$. The characterization of those functions f for which (3.1) with $\lambda = \lambda_1$ is solvable remains an open question.

Let us observe finally that this antimaximum principle as well as the above nonexistence result have been extended recently to some strictly cooperative elliptic systems involving the p-Laplacian (cf. [Fl-Go-Ta-dT$_2$]).

References

[An] A.ANANE, *Simplicité et isolation de la première valeur propre du p-laplacien avec poids*, C.R.Ac. Sci. Paris, 305 (1987), 725-728.

[An-Ts] A.ANANE et N.TSOULI, *On the second eigenvalue of the p-Laplacian*, Proc. Conf. Fès, Pitman Research Notes Math., Vol. 343, pp. 1-9, Ed. A.Benkirane et J.-P.Gossez, 1996.

[Cl-Pe] P.CLEMENT and L.PELETIER, *An antimaximum principle for second order elliptic operators*, J. Diff. Eq., 34 (1979), 218-229.

[Cu] M.CUESTA, *Le spectre de Fučik des opérateurs laplacien et p-laplacien*, Ceremath, Toulouse, 1 (1995).

[Cu-Go] M.CUESTA and J.-P.GOSSEZ, *A variational approach to nonresonance with respect to the Fučik spectrum*, Nonli. Anal., T.M.A., 19 (1992), 487-500.

[Da$_1$] N.DANCER, *On the Dirichlet problem for weakly nonlinear elliptic partial differential equations*, Proc. Royal Soc. Edimb., 76A (1977), 283-300.

[Da$_2$] N.DANCER, *Generic domain dependance for nonsmooth equations and the open set problem for jumping nonlinearities*, Proc. Royal Soc. Edimb., 1995.

[DF-Go] D.DE FIGUEIREDO and J.-P.GOSSEZ, *On the first curve of the Fučik spectrum of an elliptic operator*, Diff. Int. Equat., 7 (1994), 1285-1302.

[Fl-He-dT] J.FLECKINGER, J.HERNANDEZ and F.de THELIN, *On maximum principles and existence of positive solutions for some cooperative elliptic systems*, Diff. Int. Equat., 8 (1995), 69-85.

[Fl-Go-Ta-dT$_1$] J.FLECKINGER, J.-P.GOSSEZ, P.TAKAC et 46.de THELIN, *Existence, non-existence et principe de l'antimaximum pour le p-laplacien*, C.R.Ac. Sc. Paris, 321 (1995), 731-734.

[Fl-Go-Ta-dT$_2$] J.FLECKINGER, J.-P.GOSSEZ, P.TAKAC and 46.de THELIN, *Nonexistence of solutions and an antimaximum principle for cooperative systems with the p-Laplacian*, Math. Nachrichten, to appear.

[Fu] S.FUCIK, *Solvability of nonlinear equations and boundary value problems*, Reidel, Dordrecht, 1980.

[Gi-Tr] D.GILBARG and N.TRUDINGER, *Elliptic Partial Differential Equations of Second Order*, Springer-Verlag, 1977.

[Go-Ma] J.-P.GOSSEZ and A.MARCOS, *On the first curve in the Fučik spectrum of a mixed problem*, Proc. Conf. Trieste, M. Dekker, Ed. E. Mitidieri, to appear.

[Pr-We] M.PROTTER and H.WEINBERGER, *Maximum Principle in Differential Equations*, Prentice-Hall, 1967.

[Ta] P.TAKAC, *An abstract form of maximum and antimaximum principles of Hopf's type*, à paraître.

[Va] J.VAZQUEZ, *A strong maximum principle for some quasilinear elliptic equations*, Appl. Math. Optim., 12 (1984), 191-202.

J.-P.GOSSEZ[1]
Département de Mathématique, C.P.214
Université Libre de Bruxelles
1050 Bruxelles

[1]Partially supported by the EC grant ERBCHRXCT 940555

S I POHOZAEV

The Fibering Method in Nonlinear Variational Problems

Introduction

Let us begin with the example. Consider the following BVP

$$(0.1)\qquad \begin{cases} \Delta_p u + \lambda|u|^{p-2}u = h(x) & \text{in } \Omega \subset \mathbb{R}^N, \\ u = 0 & \text{on } \partial\Omega. \end{cases}$$

Here Δ_p is the $p-$Laplacian :

$$\Delta_p u = \operatorname{div}(|\nabla u|^{p-2}\nabla u)$$

with $p > 1, \lambda \in \mathbb{R}$ and $h \in W_{p'}^{-1}(\Omega) \equiv (\overset{\circ}{W}{}_p^1(\Omega))^*$, $\dfrac{1}{p'} + \dfrac{1}{p} = 1$, where Ω is a bounded domain in $\mathbb{R}^N$. Introduce the definition of the spectrum of the $p-$Laplacian :

$$\sigma_p = \{\lambda \in \mathbb{R} | \exists u_\lambda \in \overset{\circ}{W}{}_p^1(\Omega)$$

$$\text{with } \|u\|_{1,p} \neq 0 \text{ such that } \Delta_p u_\lambda + \lambda|u_\lambda|^{p-2}u_\lambda = 0 \text{ in } \Omega\}.$$

Note, for $p = 2$ we have the classical linear BVP and the classical definition of the spectrum for the Laplacian operator.

From the general theorem [18] on the solvability of nonlinear equations (from the relevant class) it follows the so called "Nonlinear Fredholm Alternative" for BVP (0.1).

If $\lambda \notin \sigma_p$ then for any $h \in W_{p'}^{-1}(\Omega)$ there exists a solution u of BVP (0.1).

The proof of this general theorem used the elements of the fibering method. Note, for λ greater than the first eigenvalue $\lambda_1 > 0$, the appropriate nonlinear operator $A_p : \overset{\circ}{W}{}_p^1(\Omega) \to W_{p'}^{-1}(\Omega)$ defined by

$$A_p(u) \equiv \Delta_p u + \lambda|u|^{p-2}u$$

is not coercive operator. The classical theory of nonlinear monotone coercive operators, developed by M. Vishik, F. Browder, G. Minty and other mathematicians (see for instance [15]) is not applicable to BVP (0.1) with $\lambda > \lambda_1$.

The main purpose of these lectures is to describe the approach, that may be applied to the nonlinear noncoercive variational problems.

In lectures we consider the fibering method proposed in [19] for investigation of nonlinear boundary value problems.

Let X and Y be Banach spaces, and let A be an operator (nonlinear in general) acting from X to Y.

We consider the equation

$$A(u) = h. \tag{0.2}$$

In the case when the operator is "globally nice", for example, demicontinuous, monotone, and coercive for $Y = X^*$, methods for investigating such equations have been sufficiently developed and are widely used in various problems in mathematical physics [15]. In the case when A is "locally bad", for example, not invertible in a neighborhood of some point, the Ljapunov–Schmidt method is used, based on the representation of the desired solution in the form $u = u_1 + u_2$, where u_1 is an element of some (as a rule) finite–dimensional subspace of X, and u_2 is in a corresponding complementary subspace. This approach is applicable in local nonlinear analysis of nonlinear equations and global analysis of equations close to being linear [16].

To investigate a wide class of noncoercive problems and to find solutions of such problems, wide use of the Ambrosetti–Rabinowitz method has been made, which is based on the "saddle point" lemma (Mountain Pass Lemma) [2].

Nevertheless there is a large number of nonlinear problems, where this powerful method ceases to work. Let us mention for instance the Ovsjannikov problem (1959)

$$\begin{cases} \Delta u + u^2 = h(x) & \text{in } \Omega \subset \mathbb{R}^N (N < 6), \\ u = 0 & \text{on } \partial\Omega. \end{cases} \tag{0.3}$$

The fibering method presented here has sometimes definite advantages over the known methods.

1 Description of the method

The main idea of this approach is very simple. If we can't find a "good" approach to the equation (0.2) in the original space X, we find a "good" approach in more wide space $\tilde{X}$, $X \subset \tilde{X}$, and with appropriate $\tilde{Y}$, $Y \subset \tilde{Y}$. Thus we pass from the operator $A : X \to Y$ to the new extended operator $\tilde{A} : \tilde{X} \to \tilde{Y}$ and we get new extended equation

$$\tilde{A}(\tilde{u}) = \tilde{h}$$

instead of the original equation (0.2) with $\tilde{A}|_X = A : X \to Y$.

A further development of this approach is to construct, for a given triple (X, A, Y), a corresponding triple (ξ, α, η). Here ξ and η are fibrations of the spaces X and Y respectively, and α a morphism of ξ into η. The correspondence

$$(X, A, Y) \to (\xi, \alpha, \eta)$$

is determined by the initial triple and, for given X and Y, by the operator A from X to Y.

Here we consider the simplest case of this approach, namely, we take as an extended space

$$\tilde{X} = \mathbb{R} \times X.$$

In this case we get so called one–parametric fibering method. The one–parametric fibering method is based on representation of a solution of the equation (0.2) in the form

$$u = tv. \tag{1.1}$$

Here t is a real parameter ($t \neq 0$ in some open set $J \subseteq \mathbb{R}$), and v is an element of the Banach space X ($v \neq 0$) satisfying the condition

$$H(t, v) = c \neq 0. \tag{1.2}$$

Generally speaking, any functional satisfying a sufficiently general condition (see Chapter 2) can be taken as the fibering functional $H(t, v)$.

In particular the norm $H(t, v) \equiv \|v\|$ can be taken as a fibering functional $H(t, v)$. Then the condition (1.2) takes the form $\|v\| = 1$. In this case we get a so–called spherical fibering.

Here solution $u \neq 0$ of (0.2) is sought in the form (1.1), where $t \in \mathbb{R}$ and $v \in S = \{w \in X | \|w\| = 1\}$.

A description of the method of spherical fibering and some of its applications are given in [19–23]. We remark that the elements of this method were used as far back as [18] in establishing the "Fredholm alternative" for nonlinear odd and homogeneous (mainly) strongly closed operators.

Thus, the essence of the one–parametric fibering method consists in imbedding the space X of the original problem (0.2) in the larger space $\mathbb{R} \times X$ and investigating the new problem of conditional solvability in the new space $\mathbb{R} \times X$ under condition (1.2).

This method makes possible to get both new existence theorems and new nonexistence theorems for nonlinear boundary value problems.

Further, in the investigation of solvability of boundary value problems this method makes it possible to separate the algebraic and the topological factors of the problem, which affects the number of its solutions.

2 General theory of the one–parametric Fibering method in variational problems

Let X be a real Banach space with a norm $\|w\|_X$ which is differentiable for $w \neq 0$, and let f be a functional on X of class $C^1(X \setminus \{0\})$.

We associate with the functional f a functional $\tilde{f}$ defined on $\mathbb{R} \times X$ by

$$\tilde{f}(t, v) = f(tv) \tag{2.1}$$

where $t \in \mathbb{R}$ and $v \in X$.

We consider this functional for $(t, v) \in J \times S$. Here J is an arbitrary nonempty open set in $\mathbb{R}$, and S is the unit sphere in $X : S = \{x \in X | \|x\| = 1\}$.

Theorem 2.1 *Let X be a real Banach space with the norm differentiable on $X \setminus \{0\}$, and let $(t, v) \in (J \setminus \{0\}) \times S$ be a conditional stationary point of the functional $\tilde{f}$, defined on $J \times S$. Then the vector $u = tv$, $t \neq 0$, is a stationary (critical) point of the functional f, i.e., $f'(u) = 0$.*

The proof is obvious. Indeed, at the conditional stationary point $(t, v) \in (J \setminus \{0\}) \times S$ we have that

$$\lambda \tilde{f}'_v(t, v) = \mu \|v\|', \tag{2.2}$$

$$\tilde{f}'_t(t, v) = 0, \tag{2.3}$$

with $\lambda^2 + \mu^2 \neq 0$. Here the prime and subscript mean the derivative with respect to the corresponding variable, and the derivative of the functional with respect to v is understood as the values of the derivative of the functional with respect to v in the space X for $v \in S$.

By (2.2),

$$\lambda \langle \tilde{f}'_v(t, v), v \rangle = \mu \langle \|v\|', v \rangle,$$

where (w^*, u) is the value of a functional w^* from the dual space X^* on an element u from X. Then from this and the equalities

$$\langle \tilde{f}'_v(t, v), v \rangle = t \langle f'(iv), v \rangle = t \tilde{f}'_t(t, v),$$

$$\langle \|v\|', v \rangle = 1 \text{ for } v \in S,$$

we get that $\lambda t \tilde{f}'(t, v) = \mu$.

It follows from this equality and (2.3) that $\mu = 0$. Then $\lambda \neq 0$ and by (2.2)

$$t f'(u) = \tilde{f}'_v(t, v) = 0$$

for $u = tv$, $t \neq 0$. Consequently, $f'(u) = 0$, and Theorem 2.1 is proved.

We now consider a more general fibering (decomposition). For this we introduce a fibering functional $H(t, v)$ defined on $\mathbb{R} \times X$ and we consider the functional $\tilde{f}(t, v)$ under the condition (1.2). As $H(t, v)$ we can, in general, take an arbitrary functional that is differentiable under condition (1.2) and satisfies the inequality

$$\langle H_v, v \rangle \neq t H_t \text{ for } H(t, v) = c \neq 0. \tag{2.4}$$

Theorem 2.2 *Let H be a functional of the indicated class. Let $(t, v) \in J \times X$ with $t \neq 0$ and $v \neq 0$ be a conditional critical point of the functional $\tilde{f}(t, v)$ under condition (1.2). Then the point $u = tv$ is a nonzero critical point of the original functional f, i.e., $f'(u) = 0$ and $u \neq 0$.*

The proof is analogous to the previous proof. Indeed, at the conditional critical point (t, v) we have that

$$\mu \tilde{f}'_v(t,v) = \lambda H'_v(t,v), \qquad \mu \tilde{f}'_t(t,v) = \lambda H'_t(t,v), \tag{2.5}$$

with $\mu^2 + \lambda^2 \neq 0$; on the other hand,

$$\tilde{f}'_v(t,v) = t f'_v(tv), \qquad \tilde{f}'_t(t,v) = \langle f'_t(tv), v\rangle.$$

Then from (2.5) we get that

$$\mu t f'_v(tv) = \lambda H'_v(t,v), \qquad \mu \langle f'_t(tv), v\rangle = \lambda H'_t(t,v). \tag{2.6}$$

From this we get

$$\mu t\langle f'_v(t,v), v\rangle = \lambda \langle H'_v(t,v), v\rangle,$$
$$\mu t\langle f'_v(tv), v\rangle = \lambda t H'_t(t,v).$$

Consequently,

$$\lambda\langle H'_v(t,v), v\rangle = \lambda t H'_t(t,v)$$

for $t \neq 0$ and $H(t,v) = c \neq 0$. Then, by condition (2.4), we get that $\lambda = 0$, and hence $\mu \neq 0$.

As a result, the first equation in (2.6) takes the form $f'(u) = 0$, with $u = tv \neq 0$.

3 A comparison with the Ljapunov–Schmidt approach

Now, we compare the fibering method with the classical Ljapunov–Schmidt method. We restrict ourselves to the spherical fibering method. In this case

$$H(t,v) \equiv \|v\|$$

and the condition (2.4) with $c = 1$ takes the form

$$\langle Hv, v\rangle = \|v\| = 1 \neq tH_t \equiv 0.$$

Due to the Ljapunov–Schmidt approach we seek a solution of (0.2) in the form $u = u_1 + u_2$, where u_1 is an element of a (usually finite dimensional) subspace of X and u_2 an element of a suitable, "good" complement.

Then from (0.2) we get the system of equations

$$\begin{cases} A_1(u_1, u_2) = h_1, \\ A_2(u_1, u_2) = h_2, \end{cases} \tag{3.1}$$

where the second equation for fixed $v_1 \in X_1$ is a correct, well–posed, equation and it has unique solution

$$u_2 = T(u_1, h_2).$$

Substituting this expression into the first equation we derive the so–called Ljapunov–Schmidt bifurcation equation

$$\mathcal{A}(u_1, h_2) = h_1, \tag{3.2}$$

where

$$\mathcal{A}(u_1, h_2) = A_1(u_1, T(u_1, h_2)).$$

Following the spherical fibering method we seek a solution of the variational problem

$$f'(u) = 0$$

in the form

$$u = tv,$$

where $t \in \mathbb{R}$ and $v \in S = \{u \in X | \|u\| = 1\}$.

Then the original variational problem is equivalent to the system

$$\begin{cases} \langle f'(tv), v\rangle = 0, \\ f'_\tau(tv) = 0 \qquad \text{for } v \in S. \end{cases} \tag{3.3}$$

Here f'_τ is the tangential derivative of f on the unit sphere S. The first equation is a scalar equation with respect to the scalar parameter t and with the functional parameters $v \in S$.

This equation plays the same role as the bifurcation equation in the Ljapunov–Schmidt approach.

Indeed, if we have a solution of this equation

$$t = t(v),$$

we get the induced functional $\tilde{f}(v) = f(t(v)v)$. The conditional critical point $v_c \in S$ of $\tilde{f}$ with $t_c = t(v_c) \neq 0$ generates a critical point u_c of the original functional f by formula

$$u_c = t_c v_c.$$

From the geometrical point of view the Ljapunov–Schmidt approach using the representation $u = u_1 + u_2$, i.e.

$$X = X_1 + X_2,$$

corresponds to the Cartesian coordinates.

From the geometrical point of view the spherical fibering method using the representation

$$u = tv \text{ with } \|u\| = 1,$$

corresponds to the spherical coordinates. Then it is clear that the derivative $f'(u)$

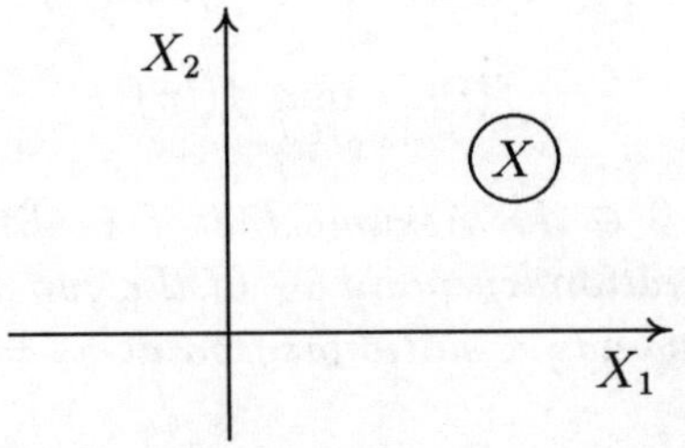

Figure 1:

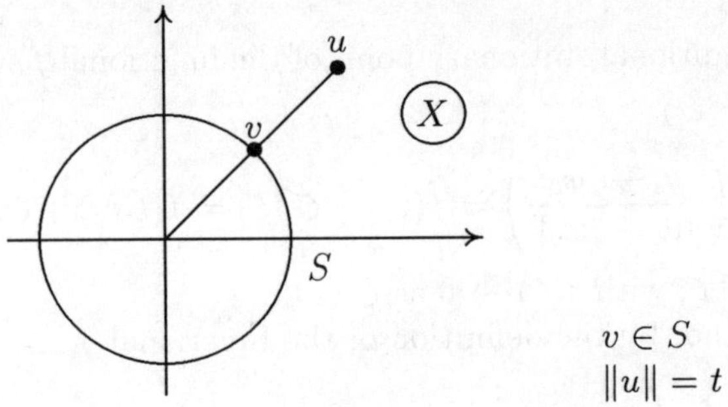

Figure 2:

is equal to 0 for $u \neq 0$ if and only if the radial derivative $\langle f_t'(tv), v\rangle = 0$ and the tangential derivative

$$f_r'(tv) = 0 \text{ for } v \in S.$$

From this geometrical point of view the introduction of the general fibering functional $H(t, v)$ corresponds to the introduction of "curvelinear" system of coordinates.

4 Minimax realization of the fibering method in variational problems

Let X be a real Banach space with norm differentiable on $X \setminus \{0\}$, let f be a functional on X belonging to the class $C^1(X \setminus \{0\})$, and let J be a nonempty open subset of $\mathbb{R}$. We now consider the functional

$$\sup_{t \in J} f(tv)$$

for $v \in S$.

Theorem 4.1 *Suppose that for any $v \in S$ the quantity*

$$\hat{f}(v) = \max_{t \in J} f(tv) \tag{4.1}$$

exists, and $f(v) > f(0)$ if $0 \in J$. Assume that f is differentiable on the sphere S. Then to each conditional stationary point v_c of the functional f, with respect to the unit sphere S, there corresponds a stationary point $u_c = t_c v_c$ of f with $t_c \in J \setminus \{0\}$ such that $f(t_c v_c) = \hat{f}(v_c)$.

Proof Assume the theorem is false, i.e. $f'(u_c) \neq 0$. Then $t_c f'(u_c) \neq 0$, since $t_c \neq 0$. Then there exists an element $w_0 \in X$ such that

$$t_c \langle f'(u_c), w_0 \rangle > 0. \tag{4.2}$$

Since $v_c \in S$ is a conditional stationary point of the functional f, which is differentiable on S, it follows that

$$\hat{f}\left(\frac{v_c + \zeta w_0}{\|v_c + \zeta w_0\|}\right) = \hat{f}(v_c) + \zeta \varepsilon(\zeta) = f(t_c v_c) + \zeta \varepsilon(\zeta) \tag{4.3}$$

for sufficiently small ζ, with $\varepsilon(\zeta) \to 0$ as $\zeta \to 0$.

On the other hand, by the definition of the functional $\hat{f}$,

$$f\left(t \frac{v_c + \zeta w_0}{\|v_c + \zeta w_0\|}\right) \leq \hat{f}\left(\frac{v_c + \zeta w_0}{\|v_c + \zeta w_0\|}\right) \qquad \forall t \in J. \tag{4.4}$$

By assumption, $\max_{t \in J} f(tv_c) = f(t_c v_c)$ is attained on the open set $J \setminus \{0\}$; hence $t_c \|v_c + \zeta w_0\| \in J \setminus \{0\}$ for sufficiently small ζ, because $\|v_c\| = 1$. Then, setting $t = t_c \|v_c + \zeta w_0\|$ in (4.4) for sufficiently small ζ, we get by (3.3) that

$$f(t_c v_c + \zeta t_c w_0) \leq f(t_c v_c) + \zeta \varepsilon(\zeta). \tag{4.5}$$

Since f is differentiable,

$$f(t_c v_c + \zeta t_c w_0) = f(t_c v_c) + \zeta t_c \langle f'(u_c), w_0 \rangle + \zeta \varepsilon_1(\zeta),$$
$$\varepsilon_1(\zeta) \to 0 \text{ as } \varepsilon \to 0.$$

Then it follows from (4.5) that

$$\zeta t_c \langle f'(u_c), w_0 \rangle \leq \zeta \varepsilon_2(\zeta), \qquad \varepsilon_2(\zeta) \to 0 \text{ as } \varepsilon \to 0.$$

Considering this inequality for sufficiently small positive ζ, we get

$$t_c \langle f'(u_c), w_0 \rangle \leq 0.$$

In view of (4.2) this inequality contradicts the assumption. Theorem 4.1 is proved. ■

Remark Let J be a nonempty open set in $\mathbb{R}$ that is symmetric with respect to $0 \in \mathbb{R}$, i.e., $t \in J$ implies that $-t \in J$.

Suppose that the functional $\hat{f}$ defined by (4.1) exists. Then it is even. This makes possible to use the Lyusternik–Schnirel'mann theory for certain functionals that are not even, and to obtain existence theorems of many geometrically different stationary points.

Example Let Ω be a bounded domain in R^N, $N \leq 3$, with smooth boundary $\partial\Omega$. Consider the following problem:

$$-\Delta u - u^3 + \left(\int_\Omega hu\, dx\right)^2 h = 0 \text{ in } \Omega,$$

$$u = 0 \text{ on } \partial\Omega.$$

By Theorem 4.1 this problem possesses a countable set of distinct solutions in $\overset{\circ}{W}{}^1_p(\Omega)$ for each $h \in W_2^{-1}(\Omega)$. Note that in this case the Euler functional

$$f(u) = \frac{1}{2}\int_\Omega |Du|^2 dx - \frac{1}{4}\int_\Omega u^4 dx + \frac{1}{3}\left(\int_\Omega hu\, dx\right)^3$$

is not even.

□

5 On the connection between conditional critical points and critical points of functionals

In the previous sections the fibering method was used to establish a connection between conditional critical points and critical points of functionals. We now consider this connection from a somewhat different point of view.

Let l be a differentiable mapping (in particular, it can also be a constant) from a real Banach space X into X. With a functional f twice differentiable on X we associate a functional f_l by the formula

$$f_l(u) = \langle f'(u), l(u)\rangle.$$

It is clear that a critical point u of f satisfies the condition $f_l(u) = 0$ and hence it is a conditional critical point of f, considered under the constraint $f_l(u) = 0$. The following simple theorem contains conditions for the validity of the converse.

Theorem 5.1 *Let f be a twice differentiable (in the Gâteaux sense) functional on a real Banach space X. Suppose that there exists a differentiable mapping l from X to X such that at a conditional critical point u_0 of the functional f, considered on the constraint $f_l(u) = 0$, the inequality*

(5.1) $$\langle f_l'(u_0), l(u_0)\rangle \neq 0$$

holds for $f_l(u_0) = 0$. *Then the conditional critical point* u_0 *of* f *is an unconditional critical point of* f.

The proof is obvious. Indeed, at u_0 we have

$$\lambda f'(u_0) = \mu f_l'(u_0), \qquad \lambda^2 + \mu^2 \neq 0. \tag{5.2}$$

Hence

$$\lambda\langle f'(u_0), l(u_0)\rangle = \mu\langle f_l'(u_0), l(u_0)\rangle \text{ with } \langle f'(u_0), l(u_0)\rangle = f_l(u_0) = 0.$$

Using (5.1) we get that $\mu = 0$, and hence $\lambda \neq 0$. Then (5.2) implies that $f'(u_0) = 0$. Theorem 5.1 is proved.

6 The choice of the fibering functional

Realization of the fibering method depends on the choice of the fibering functional $H(t, v)$ satisfying condition (2.4). As the simplest functional we can take the norm of the Banach space under the condition that it is differentiable away from the origin. However, this choice is not unique.

Here we propose as a fibering functional a functional naturally generated by the problem itself, i.e., the functional f. We remark, however, that this choice is not necessary.

The scalar equation

$$\langle f'(tv), v\rangle = 0, \tag{6.1}$$

of the scalar parameter $t = t(v)$ is the defining bifurcation equation in the fibering method.

To separate algebraically different solutions of this equation it seems to be natural to take as a fibering functional H (in the case of a functional f of class $C^3(X \setminus \{0\})$):

$$H(t, v) \equiv \langle f''(tv)v, v\rangle. \tag{6.2}$$

The standard notation is used here and below.

In this case the relation

$$\langle H_v, v\rangle - tH_t = 2H, \tag{6.3}$$

is valid for the functional H. The proof uses the following equalities:

$$\langle H_v, v\rangle = \frac{d}{d\zeta}H(t, \zeta v)|_{\zeta=1}$$

and

$$tH_t = \frac{d}{d\zeta}H(\zeta t, v)|_{\zeta=1}.$$

Then

$$H(t, \zeta v) = \zeta^2 \langle f''(\zeta t v)v, v\rangle$$

and

$$\langle H_v, v\rangle = \frac{d}{d\zeta} H(t, \zeta v)|_{\zeta=1}$$

$$= 2\langle f''(tv), v, v\rangle + \frac{d}{d\zeta}\langle f''(\zeta tv)v, v\rangle|_{\zeta=1} = 2H + tH_t.$$

By (6.3), the condition (2.4) for nondegeneracy of the fibering functional H defined by (6.2) turns out to hold always. When such a fibering functional is chosen, the solution $t = t(v)$ of (6.1) a priori inherits the smoothness (with a loss of one derivative) of the original functional on this set, if it exists.

Accordingly, the problem of finding critical points (different from zero in X) of the functional $f \in C^3(X \setminus \{0\})$ reduces to the problem of finding conditional critical points of the functional $f(tv)$ under the constraint

$$\langle f''(tv)v, v\rangle = c \neq 0, \qquad t \in J.$$

The substitution $v \to v_1 = kv$ $(k \neq 0)$ reduces this problem to the problem of finding conditional critical points of $f(tv)$ under the constraint

$$\langle f''(tv), v, v\rangle = c_0 \text{ and } t \in I_1. \tag{6.4}$$

Here the scalar parameter c_0 takes either the value $+1$ or the value -1, and I_1 is the corresponding image ($t \to t_l = \frac{t}{k}$) of the original nonempty open set J. In this problem the previous notation t and v is kept instead of t_1 and v_1.

Theorem 6.1 *Let f be a functional on X of the class $C^3(X \setminus \{0\})$, and let $(t, v) \in I_1 \times X$ with $t \neq 0$ and $v \neq 0$ be a conditional critical point of the functional $f(tv)$ under the condition (6.4). Then the point $u = tv$ is a nonzero critical point of f.*

The proof follows immediately from Theorem 2.2.

7 A parameter–free realization of the fibering method

In the general case the fibering method reduces the original variational problem to a parametric variational problem and to the investigation of its conditional critical points. In the case when the fibering functional H of the original variational problem is defined by (6.2) it is possible to eliminate the parameter t in the new variational problem. Indeed, the condition (6.4) for a functional f of class $C^3(X \setminus \{0\}$, means that

$$\frac{d}{dt}\langle f'(tv), v\rangle = \langle f''(tv)v, v\rangle \neq 0$$

on the set (6.4). Thus, condition (6.4) for a functional f of class $C^3(X\setminus\{0\})$ enables us to single out in the defining equation (6.1) the algebraically different smooth solutions $t_i(v)$ $(i = 1,\dots,N)$ (if they exist), and to eliminate the scalar parameter t.

The problem of finding critical points (different from zero in X) of a functional $f \in C^3(X \setminus \{0\})$ reduces to the problem of finding conditional critical points of the functional

$$(7.1) \qquad \begin{cases} F_i(v) = f(t_i(v)v) \\ \text{under the corresponding conditions} \\ H_i(v) \equiv \langle f''(t_i(v)v))v, v\rangle = c_0 \end{cases}$$

with $c_0 \pm 1$. Here $t_i(v)$ is the corresponding solution of (6.1) $(i = 1,\dots,N)$.

Theorem 7.1 *Let f be a functional defined on X of class $C^3(X \setminus \{0\})$, let $t_i(v)$ be the solution of (6.1) under constraint (6.4), and let v_i be a conditional critical point of the problem (7.1) with $t_i(v_i) \neq 0$. Then the point $u_i = t_i(v_i)v_i$ is a nonzero critical point of f.*

The proof follows from Theorem 6.1, since the pair $(t_i, v_i) \in I_1 \times X$ satisfies the condition of this theorem.

8 A particular realization of the fibering method

From the functional point of view it is sometimes convenient in studying a variational problem to take the fibering (normalizing) functional H to be the principal part of the original functional. Then when there is a complementary weakly continuous functional, the original problem reduces to the problem of investigating conditional critical points of a continuous functional of a sphere–type surface. It is simpler to investigate a variational problem in a closed ball, because the latter is convex.

We present a class of variational problems in which this approach can be implemented. Suppose that on a Banach space X with norm differentiable away from the origin and the given functional f of class $C^1(X)$ has the form $f(u) = f_0(u) + f_1(u)$, where $f_0(u)$ is the functional generating the norm of X. For definiteness we assume that $f_0(u) = \|u\|^p$, for some $p > 1$, and that $f_1 \in C^1(X)$. Here we choose the fibering functional H to be $H(v) = \|v\|^p$, so that the condition (1.2) takes the form $\|v\| = 1$, i.e., we use the method of spherical fibering.

The functional $f(tv)$ takes the form

$$f^*(t, v) = |t|^p + f_1(t, v) \text{ for } v \in S,$$

and the equation

$$(8.1) \qquad f_0'(u) + f_1'(u) = 0$$

is then equivalent to the system

$$(8.2) \qquad p|t|^{p-1}\operatorname{sgn} t + \langle f_1'(tv), v\rangle = 0, \qquad t \neq 0,$$

$$(8.3) \qquad t f_1'(tv) = \nu \|v\|', \quad v \in S.$$

Here the prime denotes the derivative of the corresponding functional.

Note that here the Lagrange multiplier ν differs from the corresponding Lagrange multiplier λ in the previous sections, where another realization of the fibering method was considered. Since the functional $\langle f_1'(tv), v\rangle$ is defined for all $v \in X$, the first scalar equation (8.2) in t can be considered for $v \in B = \{w \in X | \|w\| \leq 1\}$. Suppose that this equation has solutions $t_i(w)$ ($i = 1, \ldots, N$). Let

$$F_i(w) = |t_i|^p + f_1(t_i(w)w)$$

and consider these functionals on the closed unit ball B.

Definition 8.1 *A point $w \in B$ is called a* critical point *of the differentiable functional $F_i(w)$ in the closed unit ball B if one of the following two conditions holds: either w lies in the interior of B and it is a critical point of F_i; or w lies on the boundary $\partial B = S$ and it is a conditional critical point of the functional F on the sphere S.*

Definition 8.2 *A critical point $w \in B$ of the differentiable functional $F(w)$ is said to be a* regular *critical point if $w_i \neq 0$, $t_i(w_i) \neq 0$, and the functional t_i is differentiable at w_i. Here $t_i(w)$ is a solution of (8.2) for $v = w \in B$.*

Theorem 8.1 *Suppose that the functional $F_i(w)$ has a regular critical point $w_i \in B$. Then $w_i \in S$, and $u_i = t_i(w_i)w_i$ is a nonzero solution of (8.1).*

Proof Suppose that the regular critical point $w_i \in B$ of F_i is in the interior of the unit ball B. We study the behavior of F_i along the ray ζw_i as $\zeta \to 1$. By the differentiability of f and the differentiability of the functional f_i at the point w_i,

$$\left.\frac{dF_i(\zeta w_i)}{d\zeta}\right|_{\zeta=1} = \left[p|t_i|^{p-1} \operatorname{sgn} t_i + \langle f_1'(t_i w_i), w_i\rangle\right] \left.\frac{dt_i}{d\zeta}\right|_{d\zeta} + \langle f_1'(t_i w_i), w_i\rangle t_i.$$

Here $t_i = t_i(w_i)$. Hence, in view of (8.2) with $v = w_i$,

$$\frac{dF_i(\zeta w_i)}{d\zeta} |_{\zeta=1} = t_i \langle f_1'(t_i w_i), w_i\rangle = -p|t_i|^p \neq 0.$$

Thus a regular critical point w_i of F_i belongs to the boundary $\partial B = S$.

Then by Theorem 2.1 the point $u_i = t_i(w_i)w_i$ is a nonzero solution of (8.1).

■

Following Definitions 8.1 and 8.2, we introduce the concept of a regular extremal point $w \in B$ of the functional $F_i(w)$ replacing the word "critical" in those definitions by the word "extremal".

As a corollary of Theorem 8.1 we get the following "maximum principle".

Corollary 8.1 *Suppose that the functional $F_i(w)$ has a regular extremal point $w_i \in B$. Then $w_i \in \partial B$, and $u_i = t_i(w_i)w_i$ is a nonzero solution of (8.1).*

9 Simple examples

We consider the use of the method of spherical fibering in simple examples of known problems. In these examples the defining equation (6.1) with $v \in S$ admits an explicit smooth solution $t = t(v)$ for $v \in S$. This makes it possible to use a parameter–free realization of the method of spherical fibering (see Chapters 7 and 8).

In all the examples below Ω is a bounded domain in $\mathbb{R}^n$ with locally Lipschitz boundary $\partial\Omega$. The solutions of all problems are considered in the Sobolev space $\overset{\circ}{W}{}_2^1(\Omega)$, the dual space $(\overset{\circ}{W}{}_2^1(\Omega))^*$ of which is denoted by $\overset{\circ}{W}{}_2^{-1}(\Omega)$.

Example Consider the linear Dirichlet problem for the Poisson equation

$$\Delta u = h(x) \text{ in } \Omega, \qquad u = 0 \text{ on } \partial\Omega, \tag{9.1}$$
$$h \in W_2^{-1}(\Omega) \text{ with } \|h\|_{-1} \neq 0.$$

The following functional

$$f(u) = -\frac{1}{2}\int_\Omega |Du|^2 dx - \int_\Omega hu\ dx$$

is associated with (9.1). Following to the fibering method, we set $u = tv$.

Then the functional f takes the form

$$f(tv) = -\frac{1}{2}t^2\int_\Omega |Dv|^2 dx - t\int_\Omega hv\ dx.$$

In the spherical fibering

$$\|v\|^2 = \int_\Omega |Dv|^2 dx = 1$$

the functional $f^*(t,v)$ equals to

$$f^*(t,v) = -\frac{1}{2}t^2 - t\int_\Omega hv\ dx \tag{9.2}$$

and corresponds to the functional $f(tv)$. From the bifurcation equation

$$f_t^*(t,v) = -t - \int_\Omega hv\ dx = 0$$

we find that $t = -\int_\Omega hv\ dx$, and we substitute it in the expression (9.2) for f^*. Then

$$F(v) = f^*(t(v),v) = \frac{1}{2}\left(\int_\Omega hv\ dx\right)^2.$$

Note that the minimax realization of the fibering method

$$f^*(t,v) \to \max_{t\in R} f^*(t,v) = \frac{1}{2}\left(\int_\Omega hv\ dx\right)^2$$

gives rise to the very same functional F.

We now consider the critical points of this even functional on the unit sphere. Obviously, there exists an infinite set of conditional critical points of F on the unit sphere. In this set there are only two regular conditional critical points, v_1 and $v_2 = -v_1$, i.e., such that (in this case) $t_1 = t(v_1) \neq 0$ and $t_1 = t(v_2) \neq 0$. These are the points at which $F(v)$ attains a maximum on the unit sphere.

We remark that v_1 and v_2 are clearly maximum points of F on the closed unit ball $B = \{v \in \overset{\circ}{W}{}^1_2(\Omega) | \int_\Omega |Dw|^2 dx \leq 1\}$ (in the general case see the maximum principle–Corollary 8.1). Then by Theorem 7.1, and also by Theorem 8.1, $u_1 = t_1 v_1$ and $u_2 = t_2 v_2$ are the solutions of the Dirichlet problem (8.1), i.e.

$$u_1 = -v_1 \int_\Omega h v_1 dx \text{ and } u_2 = -v_2 \int_\Omega h v_2\, dx = -v_1 \int_\Omega h v_1\, dx = u_1.$$

Thus, the two nonzero solutions coincide.

□

Remark To clarify the fibering method we verify assumptions of Theorems 7.1 and 8.1 directly in this example. Thus, $v_1 \in S$ is a maximum point of F on the unit sphere S. Then by the Lagrange rule, at this point

$$\int_\Omega h v_1\, dx\, h = -\nu \Delta v_1, \qquad v_1 \in \overset{\circ}{W}{}^1_2(\Omega).$$

From this we find for $\int_\Omega |Dv_1|^2 dx = 1$ that $\nu = (\int_\Omega h v_1 dx)^2$, and $\nu \neq 0$, because $\max_{v \in S} F(v) > 0$ for $\|h\|_{-1} \neq 0$. Then

$$h \int_\Omega h v_1 dx = -\left(\int_\Omega h v_1 dx\right)^2 \Delta v_1$$

or, setting $t_1 = -\int_\Omega h v_1 dx \neq 0$ and $u_1 = t_1 v_1$,

$$\Delta u_1 = h, \qquad u_1 \in \overset{\circ}{W}{}^1_2(\Omega),$$

i.e., u_1 is the solution of problem (9.1). We see similarly that u_2 is also solution of this problem, and that $u_1 = u_2$.

Remark We remark that here the general formula (6.2) for the fibering functional $H(t, v)$ leads also to spherical fibering. Indeed,

$$H(t, v) \equiv \frac{d^2}{dt^2} f(tv) = -\int_\Omega |Dv|^2 dx,$$

and the general condition (6.4) takes the form

$$\int_\Omega |Dv|^2 dx = 1.$$

Comment The functional

$$\tilde{f}(v) = \frac{1}{2}\left(\int_\Omega hv\ dx\right)^2$$

induced both by the bifurcation equation

$$-t - \int_\Omega hv\ dx = 0$$

and by the minimax realization is the same and even, although the original Euler functional is not even. Further, $\tilde{f}$ has (more than) a countable set of critical points on the unit sphere S. We could expect the existence of countable set of solutions to the linear Dirichlet problem (9.1). But we know now that there exists unique solution. What's the matter? The functional $\tilde{f}$ has a continuous set of critical points on S with

$$\tilde{f}'(v) = \left(\int_\Omega hv\ dx\right) h = 0,$$

belonging to the equator

$$S_0 = \left\{\int_\Omega hv\ dx\right\} \cap S.$$

For these critical points we have from the bifurcation equation

$$t(v) = -\int_\Omega hv\ dx = 0$$

and consequently

$$u = t(v)v = 0.$$

It means that these critical points generate "invisible" solutions to (9.1). But there are two "visible" critical points v_1 and $v_2 = -v_1$ of f on S, i.e. where $\tilde{f}$ attains its maximum

$$\tilde{f}(v_1) = \tilde{f}(v_2) = \max_{v \in S} \frac{1}{2}\left(\int_\Omega hv\ dx\right)^2 > 0.$$

For these "visible" critical points we have

$$t_1 = t(v_1) = -\int_\Omega hv_1\ dx$$

and

$$t_2 = t(v_2) = -\int_\Omega hv_2\ dx = \int_\Omega hv_1\ dx.$$

Then $u_2 = t_1 v_1$ and $u_2 = t_2 v_2$ coincide, $u_2 = u_1$.

So we get, *that the set of solutions of (9.1) consits of the continuous set of "invisible" solutions and of two coinciding solutions, i.e. the classical unique solution of (9.1) is a "double" solution ("double star").* Now, if we have a small perturbation and consider, for instance the problem

$$\begin{cases} \Delta u + \varepsilon |u|^\delta u = h(x) \\ u = 0 \end{cases}$$

for sufficiently small $\varepsilon > 0$ and $\delta > 0$ we get a countable set of distinct "visible" solutions to this problem.

Example Consider the eigenfunction problem

$$\Delta u + |u|^{p-1} \operatorname{sgn} u = 0 \text{ in } \Omega \subset \mathbb{R}^n, \quad u = 0 \text{ for } \partial\Omega. \tag{9.3}$$

Here $2 < p < \dfrac{2n}{(n-2)}$, for $n > 2$ and $2 < p < \infty$ for $n = 1, 2$. The original functional f has the form

$$f(u) = \frac{1}{2}\int_\Omega |Du|^2 dx - \frac{1}{p}\int_\Omega |u|^p dx'.$$

According to the fibering method, we set $u = tv$. Then the functional f takes the form

$$f(tv) = \frac{1}{2}t^2 \int_\Omega |Du|^2 dx - \frac{|t|^p}{p}\int_\Omega |v|^p dx.$$

In the spherical fibering

$$\|v\| = \left(\int_\Omega |Du|^2 dx\right)^{\frac{1}{2}} = 1$$

the functional

$$f^*(t, v) = \frac{1}{2}t^2 - \frac{|t|^p}{p}\int_\Omega |v|^p dx$$

corresponds to the functional f (in this example). From the bifurcation equation

$$f_t^*(t, v) \equiv t - |t|^{p-1} \operatorname{sgn} t \int_\Omega |v|^p dx = 0$$

we find explicitly the real nonzero solutions

$$t = \pm\left(\int_\Omega |v|^p dx\right)^{\frac{1}{2-p}}.$$

Then the functional $F(v) = f^*(t(v), v)$ takes the form

$$F(v) = \frac{p-2}{2p}\frac{1}{(\int_\Omega |v|^p dx)^{\frac{2}{p-2}}}.$$

On the unit sphere $S \subset \overset{\circ}{W}{}^1_2(\Omega)$, we can apply the well–known Lyusternik–Shnirel'man theory, in view of which F has a countable set of geometrically different conditional critical points $v_1, v_2, \ldots$ on S, with $\|v_m\| = 1$ and $\int_\Omega |v_m|^p dx \to 0$ as $m \to \infty$. Then by Theorem 6.1, and also by Theorem 8.1, we get that problem (9.3) has a countable set of geometrically distinct solutions $\pm u_1, \pm u_2, \ldots, \pm u_m, \ldots$ with

$$u_m(x) = \frac{v_m(x)}{(\int_\Omega |v_m|^p dx)^{\frac{1}{p-2}}}$$

and $\|u_m\| \to \infty$ as $m \to \infty$.

□

Comment If we start from the astrophysical meaning of the Emden–Fowler equation (9.3) and consider a solution u_m as a "star" in the Sobolev space $\overset{\circ}{W}{}_2^1(\Omega)$ then the set of all solutions of (9.3) looks like an extending "Universe". Indeed, for any $R > 0$ there exists "a star" u_m with $\|u_m\| > R$. It follows from the limit relation

$$\|u_m\| \to \infty \text{ as } m \to \infty.$$

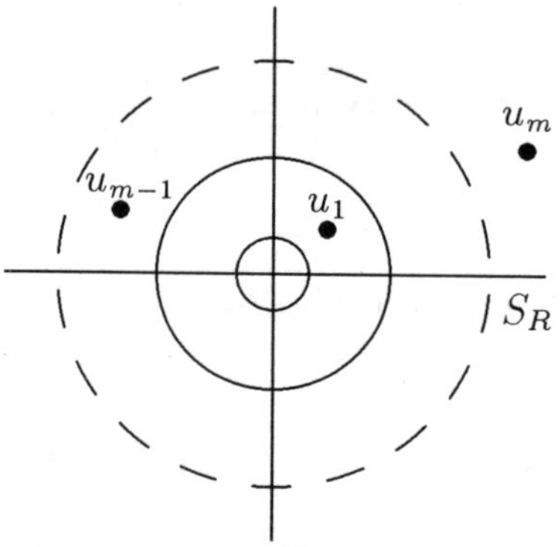

Figure 3:

10 The problem of nontrivial solutions

As in Section 9, let Ω be a bounded domain in $\mathbb{R}^n$ with locally Lipschitz boundary $\partial\Omega$. We consider the question of the existence of nontrivial solutions of the boundary value problem

(10.1) $$\Delta u + g_1(x,u)u = 0 \text{ in } \Omega \subset \mathbb{R}^n, \quad u = 0 \text{ on } \partial\Omega.$$

The conditions on the function g_1 are as follows:

(C1) The function $g_1(x,0) \equiv 0$, and g_1 is a Carathéodory function on $\Omega \times \mathbb{R}$, i.e. it is measurable with respect to x for all $u \in \mathbb{R}$ and it is continuous with respect to u for almost all $x \in \Omega$.

(C2) For $n \geq 2$ there exist positive constants A and B such that

$$|g_1(x,u)| \leq A + B|u|^m \quad \forall x \in \Omega, \ \forall u \in \mathbb{R},$$

where $0 \leq m < \dfrac{4}{(n-2)}$ for $n > 2$; and

$$|g_1(x,u)| \leq A + Be^{|u|^\alpha} \quad \forall x \in \Omega, \ \forall u \in \mathbb{R},$$

where $0 \leq \alpha < 2$ for $n = 2$.

(C3) For any function $v \in \overset{\circ}{W}{}_2^1 (\Omega)$ with $\int_\Omega |Dv|^2 dx = 1$ the equation

$$\int_\Omega g_1(x, tv(x))v^2(x)dx = 1 \tag{10.2}$$

in $t \in \mathbb{R}$ has a solution $t = t(v)$, and $t \in C^1(S)$. Here $S = \{v \in \overset{\circ}{W}{}_2^1 (\Omega) | \int_\Omega |Dv|^2 \, dx = 1\}$. Let $t = t(v)$ be solution of class $C^1(S)$. We introduce a functional

$$F(v) = -\frac{t^2(v)}{2} + \int_\Omega G(x, t(v)v)dx, \tag{10.3}$$

where $G(x, s) = \int_0^s g_1(x, y)y \, dy$.

Theorem 10.1 *Assume conditions (C1), (C2) and (C3) hold. Suppose that the weakly continuous functional F in (10.3) has a conditional critical point v on the sphere S. Then $u = t(v)v$ is a nontrivial solution of problem (10.1).*

The proof follows directly from Theorem 8.1, since the regularity of a conditional critical point of F in (10.3) on S follows from conditions (C1) and (C3).

Example Consider the boundary value problem

$$\Delta u + a(x)u^2 + u^3 = 0 \text{ in } \Omega \subset \mathbb{R}^n, \quad n \leq 3, \quad u = 0 \text{ on } \partial\Omega,$$

with $a \in L_q(\Omega)$, where $q = 1$ for $n = 1$, $q > 1$ for $n = 2$, and $q > 2$ for $n = 3$. Then by Theorem 10.1 this boundary value problem has a nontrivial solution $u \in \overset{\circ}{W}{}_2^1 (\Omega)$.

Let us now consider the equation (10.2). If to $v_1 \in S$ there corresponds a solution $t_1 = t(v_1)$, then to the function $v_2 = -v_1$ there corresponds the solution $t_2 = t(-v_1) = -t(v_1) = -t_1$. The functional F in (10.3) is even on such an odd solution. If this even functional is smooth, then the Lyusternik–Schnirel'mann theory can be applied.

□

Theorem 10.2 *Assume conditions (C1), (C2) and (C3) hold. Suppose that even functional F in (10.3) satisfies on S the Lyusternik–Schnirel'mann conditions, in any of the variants of the Lyusternik–Schnirel'mann theory. Then the boundary-value problem (10.1) has a countable set of geometrically distinct solutions.*

Proof The existence of a countable set of geometrically distinct conditional critical points of the even weakly continuous functional F on the unit sphere S follows from the Lyusternik–Schnirel'mann theory in the variant corresponding to the condition adopted.

The regularity of each conditional critical point of the functional on the sphere S follows from conditions (C1) and (C3), since a solution of (10.2) at a conditional critical point of the functional on S is nonzero and differentiable at this point. Then we get Theorem 10.2 from Theorem 8.1.

■

Example Consider the boundary value problem

$$\Delta u + \lambda(x)u + \nu(x)|u|^m \operatorname{sgn} u = 0 \text{ in } \Omega \subset \mathbb{R}^n, \quad u = 0 \text{ on } \partial\Omega.$$

Here $1 < m < \dfrac{(n+2)}{(n-2)}$ for $n > 2$ and $1 < m$ for $n = 1, 2$. Denote by λ_0 the first eigenvalue of the Laplace operator in the domain Ω with homogeneous Dirichlet boundary condition.

We get by Theorem 10.2 that for any functions λ, $\mu \in C(\overline{\Omega})$ with $\lambda(x) < \lambda_0$ and $\mu(x) > 0$ in $\overline{\Omega}$ this problem has a countable set of geometrically distinct solutions in the Sobolev space $\overset{\circ}{W}{}_2^1(\Omega)$.

□

Example We consider the boundary value problem

$$\begin{cases} \Delta u + a(x)|u|^\alpha \operatorname{sgn} u + u^3 = 0 & \text{in } \Omega \subset \mathbb{R}^n, \quad n \leq 3, \\ u = 0 & \text{on } \partial\Omega. \end{cases}$$

Here $1 < \alpha < 3$ and $a \in L_q(\Omega)$, where $q = 1$ for $n = 1$, $q > 1$ for $n = 2$, and $q > \dfrac{6}{5-\alpha}$ for $n = 3$. Then by Theorem 10.2 this problem has a countable set of geometrically distinct solutions in the Sobolev space $\overset{\circ}{W}{}_2^1(\Omega)$.

□

11 A boundary value problem with even nonlinearity

Here on the basis of Theorem 8.1 we consider the following boundary value problem in a bounded domain $\Omega \subset \mathbb{R}^n$ with $n \leq 5$ and with smooth boundary $\partial\Omega$:

$$\Delta\Phi + \Phi^2 = \varphi(x) \text{ in } \Omega, \quad u = h(x) \text{ on } \partial\Omega, \tag{11.1}$$

where $\varphi \in W_2^{-1}(\Omega) = (\overset{\circ}{W}{}_2^1(\Omega))^*$ and $h \in W_2^{\frac{1}{2}}(\partial\Omega)$.

Let H be a harmonic function in $W_2^1(\Omega)$ such that $\Delta H = 0$ in Ω and $H = h(x)$ on $\partial\Omega$. Then the original boundary value problem is equivalent ($\Phi = u + H$) to the following problem:

$$\Delta u + (u + H(x))^2 = \varphi(x) \text{ in } \Omega, \quad u = 0 \text{ on } \partial\Omega. \tag{11.2}$$

The following functionals on the Sobolev space $\overset{\circ}{W}{}_2^1(\Omega)$ are associated with (11.2):

$$f(u) = \int_\Omega |Dv|^2 dx = \|u\|^2,$$

$$g(u) = \int_\Omega \left[-\frac{2}{3}(u+H)^3 + 2\varphi u + \frac{2}{3}H^3\right] dx.$$

The bifurcation equation for $t(v)$ in this case takes the form

$$-t^2 \int_\Omega v^3\, dx + \left(1 - 2\int_\Omega Hv^2\, dx\right) t + \int_\Omega (\varphi - H^2) v\, dx = 0.$$

From this equation we get

$$\begin{aligned} t_i(v) = \Bigg\{ 1 - 2\int_\Omega Hv^2 dx + \nu_i \Bigg[\left(1 - 2\int_\Omega Hv^2 dx\right)^2 + \\ +4\int_\Omega v^3 dx \int_\Omega (\varphi - H^2) v\, dx \Bigg]^{\frac{1}{2}} \Bigg\} \left(2\int_\Omega v^3 dx\right)^{-1}, \end{aligned} \tag{11.3}$$

$$\begin{aligned} \nu_i &= \begin{cases} -1 & \text{for } i = 1, \\ 1 & \text{for } i = 2, \end{cases} \\ F_i(v) &= \frac{1}{2}\left(1 - 2\int_\Omega Hv^2 dx\right) t_i^2(v) + \frac{4}{3} t_i(v) \int_\Omega (\varphi - H^2) v\, dx \end{aligned} \tag{11.4}$$

for $\int_\Omega v^3 dx \neq 0$.

The functional F_1 is defined for all $v \in B_1 = \{w \in \overset{\circ}{W}{}_2^1(\Omega) \mid \|w\| \leq 1\}$, and F_2 is defined for $v \in B_1 \setminus \{v \mid \int_\Omega v^3 dx \neq 0\}$.

Let us consider the behavior of these functionals in the unit ball B_1 as $\|v\| \to 0$ for $\int_\Omega v^3 dx \neq 0$. Then

$$\int_\Omega v^3 dx \to 0, \quad \int_\Omega Hv^2 dx \to 0, \quad \int_\Omega (\varphi - H^2) v\, dx \to 0,$$

and

$$t_i(v) = \frac{1 - 2\int_\Omega Hv^2 dx}{2\int_\Omega v^3 dx}(1 + \nu_i) + \nu_i \left[\frac{\int_\Omega (\varphi - H^2) v\, dx}{1 - 2\int_\Omega Hv^2 dx} - \frac{(\int_\Omega (\varphi - H^2) v\, dx)^2 \int_\Omega v^3 dx}{(1 - 2\int_\Omega Hv^2 dx)^3} \right],$$

$$\begin{aligned} F_i(v) = &\left[\frac{(1 - 2\int_\Omega Hv^2 dx)^3}{6(\int_\Omega v^3 dx)^2} + \frac{\int_\Omega (\varphi - H^2) v\, dx (1 - 2\int_\Omega Hv^2 dx)}{\int_\Omega v^3 dx} \right] (1 + \nu_i) + \\ &+\nu_i \left[\frac{(\int_\Omega (\varphi - H^2) v\, dx)^2}{1 - 2\int_\Omega Hv^2 dx} - \frac{4}{3} \frac{\int_\Omega (\varphi - H)^2 v\, dx)^3 \int_\Omega v^3 dx}{(1 - 2\int_\Omega Hv^2 dx)^3} + \ldots \right] \end{aligned}$$

for $\int_\Omega v^3 \neq 0$ and $\int_\Omega Hv^2 dx \neq \frac{1}{2}$.

Concerning the boundary function $h \in W_2^{\frac{1}{2}}(\partial\Omega)$ we assume also that there exists a constant $c_0 > 0$ such that the corresponding harmonic function $H(x)$ satisfies the inequality

$$1 - 2\int_\Omega Hv^2 dx \geq c_0 \qquad \forall v \in B_1.$$

This holds, in particular, for $H(x) \leq 0$.

Concerning the function $\varphi \in W_2^{-1}(\Omega)$ we assume that there exists a constant $c_1 > 0$ such that

$$\left(1 - 2\int_\Omega Hv^2dx\right)^2 + 4\int_\Omega v^3dx \int_\Omega (\varphi - H^2)v\, dx \geq c_1 \quad \forall v \in B_1.$$

This holds, in particular, if the function $\varphi - H^2$ is sufficiently small in the norm of the dual space $W_2^{-1}(\Omega)$. Under our assumptions,

$$\sup_{v \in B_1} F_2(v) = \sup_{v \in S} F_2(v) = +\infty,$$

$$\inf_{v \in B_1} F_2(v) > -\infty, \qquad \inf_{v \in B_1} F_1(v) > -\infty.$$

Note that in the case when $\varphi(x) = H^2(x)$ a.e. in Ω the trivial solution is one of the solutions of problem (11.2). Therefore, it is assumed below that

$$\|\varphi(x) - H^2(x)\|_{W_2^{-1}(\Omega)} \neq 0.$$

Then

$$\inf_{v \in B_1} F_1(v) < 0.$$

Further, the corresponding minimum points v_1 and v_2 exist for the functionals F_1 and F_2 in the unit ball B_1 of the space $\overset{\circ}{W}{}_2^1(\Omega)$.

For the functional F_1 at the point $v_i \in B_1$ we have

$$F_1(v_1) = \min_{v \in B_1} F_1(v) < 0.$$

Then we get from the representation (11.3) for F_1 that $v_1 \neq 0$ and $t_1(v_1) \neq 0$. The functional F_1 is differentiable at v_1. Thus, the conditions of Theorem 8.1 are satisfied for F_1, and hence $u_1 = t_1(v_1)v_1$ is a solution of the boundary value problem (11.2) under the conditions on functions φ and h indicated above.

Let us now consider the functional F_2. At the point $v_2 \in B_1$ we have

$$F_2(v_2) = \min_{v \in B} F_2(v) > -\infty.$$

Then we get from the representation (11.3), (11.4) that $t_2 v_2 \neq 0$ and $t_2(v_2) \neq 0$. The functional F_2 is differentiable at the point v_2, and $\int_\Omega v_2^3 dx \neq 0$.

Thus, the conditions of Theorem 8.1 hold also for F_2, and hence $u_2 = t_2(v_2)v_2$ is a solution of (11.2) under the indicated conditions on φ and h.

Note that the solutions u_1 and u_2 are different. Indeed, if $u_1 = u_2$, then $t_1 v_1 = t_2 v_2$. From this it would follow for $v_2, v_2 \in S$ that $v_1 = \pm v_2$ and $|t_1| = |t_2|$. The last equalities contradict (10.3).

12 Applications to the local analysis

Above we have considered some applications to the global analysis of certain nonlinear BVP's. It is clear that the fibering method can be applied to the local analysis of certain nonlinear variational problems as well. We restrict ourselves to a simple example.

Let's consider the following BVP

$$\begin{cases} \Delta u + u^3 = h(x) & \text{in } \Omega, \\ u = 0 & \text{on } \partial\Omega, \end{cases} \tag{12.1}$$

where Ω is a bounded domain in $\mathbb{R}^n$ and $n < 4$.

To this BVP there corresponds the Euler functional

$$E(u) = -\frac{1}{2}\int_\Omega |Du|^2 dx - \frac{1}{4}\int_\Omega u^4 dx - \int_\Omega hu\ dx.$$

Due to the spherical fibering we have

$$u(x) = tv(x)$$

with

$$\|v\|^2 = \int_\Omega |Du|^2 dx = 1 \text{ for } v \in \overset{\circ}{W}{}_2^1(\Omega).$$

Then the Euler functional E generates

$$\tilde{E}(fv) = -\frac{1}{2}t^2 + \frac{1}{4}t^4 \int_\Omega v^4 dx - t\int_\Omega hv\ dx.$$

The bifurcation equation takes the form

$$\frac{d\tilde{E}}{dt} \equiv -t + t^3 \int_\Omega v^4 dx - \int_\Omega hv\ dx = 0.$$

It is clear (see fig. 4) that if

$$\sup_{v\in S}\left\{\left|\int_\Omega hv\ dx\right|\left(\int_\Omega |v|^4 dx\right)^{\frac{1}{2}}\right\} < \frac{2}{3\sqrt{3}}$$

then the bifurcation equation possesses three isolated smooth branches of solutions: $t_1 = t_1(v,h)$, $t_2 = t_2(v,h)$ and $t_3 = t_3(v,h)$. Substituting them in $\tilde{E}$ we get three induced functionals

$$\begin{aligned} \tilde{f}_i(v) &= \tilde{E}(t_i(v,h), v) = \\ &= -\frac{1}{2}t_i^2(v,h) + \frac{1}{4}t_i^4(v,h)\int_\Omega v^4 dx - t_1(v,h)\int_\Omega hv\ dx \end{aligned}$$

$i = 1, 2, 3$. These functionals are distinct and smooth on S.

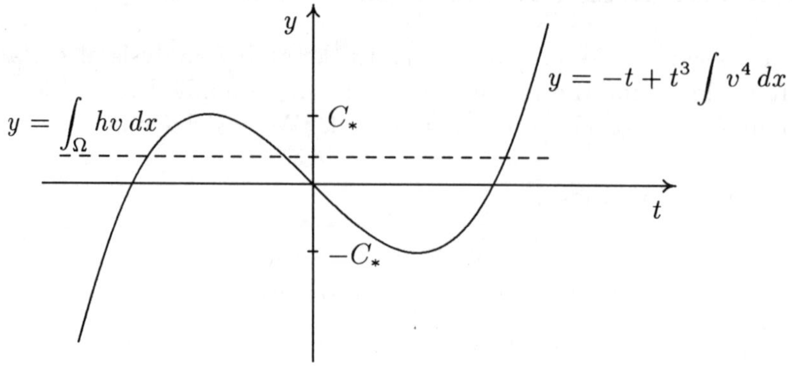

$$C_* = 2 \cdot 3^{-3/2} \left(\int_\Omega v^4 \, dx \right)^{-1/2}$$

Figure 4:

Every functional $\tilde{f}_i(v)$ $(i = 1, 2, 3)$ has a critical point v_i on S. Then the original Euler functional E (under condition on h) possesses three distinct critical points, i.e. solutions of (12.1)

$$u_1(x) = t_1 v_1(x), \quad u_2(x) = t_2 v_2(x), \quad u_3(x) = t_3 v_3(x)$$

from $\overset{\circ}{W}{}_2^1(\Omega)$ such that

$$\int_\Omega h u_1 dx \leq 0, \quad \int_\Omega h u_2 dx \leq 0$$

and

$$\int_\Omega h u_3 dx \geq 0, \quad \|u_1\| < \|u_2\|.$$

Comment For sufficiently small h the existence of solutions to (12.1) can be proved by the contraction mapping principle. The existence of the second solution can be proved by means of the existence of nontrivial solution for the difference solutions.

But, we do not know, how is it possible to prove the existence of the third distinct solution using another method.

13 Positive solutions for the p–Laplacian

Now we consider the application of the fibering method to the problem of existence of the positive solutions of the p–Laplacian. Here we follow the paper [12] by P. Drábek and S.I. Pohozaev, where additionally the existence and nonexistence problems are examined in the whole of $\mathbb{R}^N$.

We study the existence and multiplicity of positive solutions of the equation

$$- \Delta_p u = \lambda g(x)|u|^{p-2}u + f(x)|u|^{\gamma-2}u. \tag{13.1}$$

We assume $p, \lambda, \gamma \in \mathbb{R}$, $1 < p < \gamma < p^*$, where $p^* = \dfrac{Np}{N-p}$ for $p < N$ and $p^* = \infty$ for $N \leq p$. We consider the equations in the bounded domain $\Omega \subset \mathbb{R}^N$ with the Dirichlet boundary condition

$$u = 0 \text{ on } \partial\Omega. \tag{13.2}$$

This problem is studied in connection with the corresponding eigenvalue problem for the p-Laplacian

$$- \Delta_p u = \lambda g(x)|u|^{p-2}u. \tag{13.3}$$

We concentrate on the existence and multiplicity of positive solutions to (13.1) when λ belongs to a certain neighbourhood of the first eigenvalue λ_1 of (13.3). In particular, for $\lambda \in (\lambda_1, \lambda_1 + \varepsilon)$, with some $\varepsilon > 0$, we prove the existence of at least two distinct positive solutions of (13.1).

The equation (13.1) in the whole of $\mathbb{R}^N$ is studied in papers [10], [11]. Essentially the same result as here was proved in [11] using the *bifurcation argument* [8] combined with the *critical point theory*. Here we study the problem (13.1) using the *fibering method* introduced in [19], [21] and [22]. It appears that this method yields the existence and multiplicity of positive solutions in more *explicit* and *constructive* way than the approach based on the bifurcation argument. Moreover, our approach unables us to give also *nonexistence result*.

Notations and preliminaries

Let Ω be a bounded domain in $\mathbb{R}^N$, $1 < p < \infty$. We will work in the Sobolev space $W := \overset{\circ}{W}{}_0^{1,p}(\Omega)$ equipped with the norm

$$\|u\|_W = \left(\int_\Omega |\nabla u|^p dx\right)^{\frac{1}{p}}. \tag{13.4}$$

We assume that $g(x) \in L^\infty(\Omega)$, $g^+(x) \equiv 0$ (g^+ and g^- denote the positive and the negative part of the function g, respectively, i.e. $g(x) = g^+(x) - g^-(x)$). Let $1 < p < \gamma < p^*$ and $f(x) \in L^\infty(\Omega)$. Then a function $u \in W$ is called the weak solution to the problem (13.1), (13.2) if the following integral identity

$$\int_\Omega |\nabla u|^{p-2}\nabla u \nabla v \, dx = \lambda \int_\Omega g(x)|u|^{p-2}uv \, dx + \int_\Omega f(x)|u|^{\gamma-2}uv \, dx \tag{13.5}$$

holds for every $v \in W$.

A real number λ is called an *eigenvalue* and $u \in W$, $u(x) \equiv 0$, is corresponding *eigenfunction* to the problem (13.3), (13.2) if

$$\int_\Omega |\nabla u|^{p-2}\nabla u \nabla v \, dx = \lambda \int_\Omega g(x)|u|^{p-2}uv \, dx \tag{13.6}$$

holds for every $v \in W$. The following result is now well known (see e.g. [3], [6], [14]).

Lemma 13.1 *There exists the first positive eigenvalue λ_1 of the problem (13.3), (13.2) which is characterized as the minimum of the Rayleigh quotient:*

$$\lambda_1 = \min_{\substack{u \in W \\ \int_\Omega g(x)|u|^p\,dx > 0}} \frac{\int_\Omega |\nabla u|^p dx}{\int_\Omega g(x)|u|^p dx} > 0. \tag{13.7}$$

Moreover, λ_1 is simple (i.e. all the associated first eigenfunctions u are merely constant multiples of each other), λ_1 is isolated (i.e. there are no eigenvalues less than λ_1 and no eigenvalues in some right reduced neighbourhood of λ_1). There is positive (in Ω) eigenfunction u_1 corresponding to λ_1.

We will denoty by $(.,.)_W$ the duality pairing between W^* (the dual space) and W, so that the principal part of (13.1) can be written as

$$\int_\Omega |\nabla u|^{p-2} \nabla u \nabla v \; dx = (-\Delta_p u, v)_W.$$

It follows from the continuity of the Nemytskii operator [13] and the Sobolev imbedding theorem [1] that $g \in L^\infty(\Omega)$ implies that
(g0) *the functional*

$$u \to \int_\Omega g(x)|u|^p dx$$

is weakly continuous on W.
Analougously, it follows from $f(x) \in L^\infty(\Omega)$ and $1 < \gamma < p^*$ that
(f0) *the functional*

$$u \to \int_\Omega f(x)|u|^\gamma dx$$

is weakly continuous on W.

The applications of the fibering method

We will illustrate the idea of the *fibering method* (see e.g. [19], [21], [22]) on the boundary value problem (13.1), (13.2). Let us consider the *Euler functional*

$$\mathcal{I}_\lambda(u) = \frac{1}{p}\int_\Omega |\nabla u|^p dx - \frac{\lambda}{p}\int_\Omega g(x)|u|^p dx = \frac{1}{\gamma}\int_\Omega f(x)|u|^\gamma dx \tag{13.8}$$

associated with (13.1), (13.2). It is clear that the critical point of $\mathcal{I}_\lambda$ corresponds to the weak solution of the boundary value problem (13.1), (13.2). Let us split the function $u \in W$ as follows:

$$u(x) = rv(x), \tag{13.9}$$

$r \in \mathbb{R}$, $v \in W$ and substitute (13.9) into (13.8). We get

$$\mathcal{I}_\lambda(rv) = \frac{|r|^p}{p}\int_\Omega |\nabla u|^p dx - \frac{\lambda |r|^p}{p}\int_\Omega g(x)|v|^p dx - \frac{|r|^\gamma}{\gamma}\int_\Omega f(x)|v|^\gamma dx. \tag{13.10}$$

Let $u \in W$ be the critical point of $\mathcal{I}_\lambda(u)$. Then we get the bifurcation equation

$$\frac{\partial \mathcal{I}_\lambda(rv)}{\partial r} = 0,$$

i.e.

$$|r|^{p-2}r\int_\Omega |\nabla v|^p dx - \lambda |r|^{p-2}r\int_\Omega g(x)|v|^p dx - |r|^{\gamma-2}r\int_\Omega f(x)|v|^\gamma dx = 0. \tag{13.11}$$

Looking for *nontrivial solutions (critical points)* $u \equiv 0$, we have to consider $r \neq 0$. Then it follows from (13.11) that

$$\int_\Omega |\nabla v|^p dx - \lambda \int_\Omega g(x)|v|^p dx - |r|^{\gamma-p}\int_\Omega f(x)|v|^\gamma dx = 0.$$

Under the assumption $\int_\Omega f(x)|v|^\gamma dx \neq 0$ we get from here

$$|r|^{\gamma-p} = \frac{\int_\Omega |\nabla v|^p dx - \lambda \int_\Omega g(x)|v|^p dx}{\int_\Omega f(x)|v|^\gamma dx} > 0. \tag{13.12}$$

If we calculate $|r|$ from (13.12) and substitute it into (13.10), we get

$$\tilde{\mathcal{I}}_\lambda(v) := \mathcal{I}_\lambda(r(v)v) =$$

$$= \left(\frac{1}{p} - \frac{1}{\gamma}\right) \cdot \frac{|\int_\Omega |\nabla v|^p dx - \lambda \int_\Omega g(x)|v|^p dx|^{\frac{\gamma}{\gamma-p}}}{|\int_\Omega f(x)|v|^\gamma dx|^{\frac{p}{\gamma-p}}} \operatorname{sgn}\left(\int_\Omega f(x)|v|^\gamma dx\right). \tag{13.13}$$

Lemma 13.2 *The functional $\tilde{\mathcal{I}}_\lambda$ is 0–homogeneous, i.e. for every $v \in W$, $\int_\Omega f(x)|v|^\gamma dx \neq 0$ and every $t \in \mathbb{R} \setminus \{0\}$ we have*

$$\tilde{\mathcal{I}}_\lambda(tv) = \tilde{\mathcal{I}}_\lambda(v).$$

In particular, $\tilde{\mathcal{I}}_\lambda$ is even and the Gateâux derivative of $\tilde{\mathcal{I}}_\lambda$ at the point $v \in W$, $\int_\Omega f(x)|v|^\gamma dx \neq 0$, in the direction v is zero:

$$(\tilde{\mathcal{I}}_\lambda(v), v)_W = 0.$$

Moreover, it follows from (13.13) that if $v_c \in W$ is a critical point of $\tilde{\mathcal{I}}_\lambda$ then also $|v_c|$ is a critical point of $\tilde{\mathcal{I}}_\lambda$.

Note that the proof of Lemma 13.1 is obvious. It follows from here that whenever we find some *critical point* v_c of $\tilde{\mathcal{I}}_\lambda$, we can *automatically* assume that v_c is *nonnegative* in Ω.

Lemma 13.3 *Every critical point $v_c \neq 0$, $\int_\Omega f(x)|v_c|^\gamma dx \neq 0$ of $\tilde{\mathcal{I}}_\lambda$ generates the critical point $u_c \in W$ of $\mathcal{I}_\lambda$ by the formula*

$$u_c(x) = r_c v_c(x),$$

where $r_c \neq 0$ is defined by (13.12) (where we set $v := v_c$).

Proof Let $v_c \in W$, $\int_\Omega f(x)|v_c|^\gamma dx \neq 0$ be a critical point of $\tilde{\mathcal{I}}_\lambda$. Then

$$\tilde{\mathcal{I}}'_\lambda(v_c) = 0.$$

Since $\tilde{\mathcal{I}}_\lambda = \mathcal{I}_\lambda(r(v)v)$, it follows from (13.8), (13.10)

$$\tilde{\mathcal{I}}'_\lambda(v) = \frac{\partial \tilde{\mathcal{I}}_\lambda(u)}{\partial r}\frac{dr}{dv} + \tilde{\mathcal{I}}'_\lambda(u)r,$$

where $u = rv$. But we have $\dfrac{\partial \mathcal{I}_\lambda(u)}{\partial r} = 0$ due to (13.11) and so

$$\tilde{\mathcal{I}}'_\lambda(v) = r\tilde{\mathcal{I}}'_\lambda(u).$$

In the critical point v_c we have

$$r_c\mathcal{I}'_\lambda(u_c) = \tilde{\mathcal{I}}'_\lambda(v_c) = 0$$

and hence for $r_c \neq 0$ we finally get

$$\tilde{\mathcal{I}}'_\lambda(u_c) = 0.$$

■

We will show that the critical point v_c of the homogeneous functional $\tilde{\mathcal{I}}'_\lambda$ can be found by using of the *conditional variational problem* associated with $\tilde{\mathcal{I}}'_\lambda$.

Lemma 13.4 *Let us consider the constraint* $H(v) = c$, *where the functional* $H : W \to \mathbb{R}$ *satisfies the following condition*

(13.14) $$(H'(v), v)_W \neq 0 \text{ if } H(v) = c.$$

Then every conditional critical point of the problem

(13.15) $$\operatorname{crit}\{\tilde{\mathcal{I}}'_\lambda(v); H(v) = c\}$$

is the usual ciritcal point of $\tilde{\mathcal{I}}'_\lambda(v)$.

Proof Let v_c be the conditional critical point of (13.15). Then there exist μ_i, $i = 1, 2$, such that

(13.16) $$\mu_1\tilde{\mathcal{I}}'_\lambda(v_c) = \mu_2 H'(v_c),$$

where

(13.17) $$\mu_1^2 + \mu_2^2 > 0.$$

It follows from (13.16) that

(13.18) $$\mu_1(\tilde{\mathcal{I}}'_\lambda(v_c)v_c)_W = \mu_2(H'(v_c), v_c)_W.$$

But $(\tilde{\mathcal{I}}_\lambda(v_c), v_c)_W = 0$ due to Lemma 13.2 and $(H'(v_c), v_c)_W \neq 0$ due to the assumption (13.14). Then it follows from (13.17), (13.18) that $\mu_1 \neq 0$, $\mu_2 = 0$. Hence (13.16) implies

$$\tilde{\mathcal{I}}'_\lambda(v_c) = 0.$$

■

The following *scheme* for the investigation of the solvability of (13.1) is based on previous lemmas. We will associate the functional $\tilde{\mathcal{I}}_\lambda$ to the equation (13.1). First, we will prove the existence of nonzero critical points of $\tilde{\mathcal{I}}_\lambda$ under constraint given by suitable functional H. This will be the usual critical point of $\tilde{\mathcal{I}}_\lambda$ and it will generate the critical point of the Euler functional $\mathcal{I}_\lambda$ which will coincide with the weak solution of the problem (13.1).

The case $0 \le \lambda < \lambda_1$

Let us consider the problem

$$(13.19)\qquad \begin{aligned} -\Delta_p u &= \lambda g(x)|u|^{p-2}u + f(x)|u|^{\gamma-2}u \text{ in } \Omega,\\ u &= 0 \text{ on } \partial\Omega. \end{aligned}$$

Let $\lambda_1 > 0$ be the first positive eigenvalue of (13.3), (13.2) and $u_1 = u_1(x)$ be the corresponding positive eigenfunction (see Lemma 13.1). We will assume $0 \le \lambda < \lambda_1$ in this section. Set

$$(13.20)\qquad H_\lambda(v) := \int_\Omega |\nabla v|^p dx - \lambda \int_\Omega g(x)|v|^p dx.$$

It follows from Lemma 13.1 that $H_\lambda(v) \ge 0$ for any $v \in W$ and it follows from (13.20) that

$$(H'_\lambda(v), v)_W = pH_\lambda(v), \quad v \in W.$$

Therefore (13.14) is fulfilled if we consider

$$H_\lambda(v) = 1$$

as a constraint. It follows from (13.12) that we have to consider the critical points of $\tilde{\mathcal{I}}_\lambda(v)$ satisfying $\int_\Omega f(x)|v|^\gamma dx > 0$. So the following hypothesis concerning $f = f(x)$ is a natural one:

$(f1)$ $\qquad f^+(x) \not\equiv 0.$

Let us consider the conditional variational problem:

(P_λ) $\qquad$ *Find maximizer $v_c \in W$ of the problem*

$$0 < M_\lambda = \sup\left\{\int_\Omega f(x)|v|^\gamma dx;\, H_\lambda(v) = 1\right\}.$$

Proposition 13.1 *Assume $(g0)$, $(f0)$, $(f1)$. Then the problem (P_λ) has a nonegative solution.*

Proof Let us consider the set

$$W_\lambda = \{v \in W;\; H_\lambda(v) = 1\}.$$

From $(f1)$ it follows that $W_\lambda \neq \emptyset$. Next, we prove that the set is bounded in W. Due to the variational characterization (13.7) of λ_1 we get for any $v \in W_\lambda$:

$$\int_\Omega |\nabla v|^p dx = \lambda \int_\Omega g(x)|v|^p dx + 1 \leq \frac{\lambda}{\lambda_1} \int_\Omega |\nabla v|^p dx + 1 \tag{13.21}$$

for $\lambda \geq 0$. It follows from (13.21) that for $0 \leq \lambda < \lambda_1$, $v \in W_\lambda$:

$$\int_\Omega |\Delta v|^p dx \leq \frac{\lambda_1}{\lambda_1 - \lambda}.$$

Hence the maximizing sequence $\{v_n\}_{n=1}^\infty$ for (P_λ) is bounded in W. Consequently we may assume that

$$v_n \rightharpoonup v_c \text{ in } W.$$

Due to $(f0)$ we have

$$\int_\Omega f(x)|v_n|^\gamma dx \leq \int_\Omega f(x)|v_c|^\gamma dx = M_\lambda > 0. \tag{13.22}$$

Moreover, we have $H_\lambda(v_n) = 1$ and due to the weak lower semicontinuity of the norm $\|.\|_W$ and $(g0)$ we get

$$\int_\Omega |\nabla v_c|^p dx \leq \liminf_{n\to\infty} \int_\Omega |\nabla v_n|^p dx,$$
$$\int_\Omega g(x)|v_c|^p dx \leq \lim_{n\to\infty} \int_\Omega g(x)|v_n|^p dx.$$

Hence

$$H_\lambda(v_c) := \int_\Omega |\nabla v_c|^p dx \leq - \int_\Omega |v_c|^p dx \leq 1. \tag{13.23}$$

It follows from (13.22) that $v_c \not\equiv 0$ and we may also assume $v_c \geq 0$ (cf. Lemma 13.2). We prove that, actually, equality holds in (13.23). Assume that this is not true, i.e.

$$H_\lambda(v_c) < 1.$$

We find $k_c > 1$ such that

$$H_\lambda(k_c v_c) = 1.$$

Then $\tilde{v}_c = k_c v_c \in W_\lambda$ and

$$\int_\Omega f(x)|\tilde{v}_c|^\gamma dx = k_c^\gamma \int_\Omega f(x)|v_n|^\gamma dx = k_c^\gamma M_\gamma > M_\lambda,$$

which is a contradiction. Hence $v_c \in W_\lambda$ is the solution of (P_λ).

■

Theorem 13.1 *Let $1 < p < \gamma < p^*$, $0 \leq \lambda < \lambda_1$, $g, f \in L^\infty(\Omega)$ and $(f1)$ be satisfied. Then the boundary value problem (4.1) has at least one positive weak solution $u \in W \cap L^\infty(\Omega)$. Moreover $u \in C^{1,\alpha}_{\text{loc}}(\Omega)$.*

Proof Recall that $(g0)$, $(f0)$ hold under assumptions of theorem. Then due to Proposition 13.5 there is nonnegative solution v_c of the problem (P_λ). It is the critical point of $\tilde{\mathcal{I}}_\lambda$ due to Lemma 13.4 and so we can construct the critical point $u \in W$ of $\mathcal{I}_\lambda$ according to Lemma 13.3 (with $v := v_c$) and set

$$u(x) := r_c v_c(x).$$

Then $u \in W$ is the weak solution of (4.1) and $u \geq 0$ in Ω. Now, following the bootstrap argument (used e.g. in [9]) we prove $u \in L^\infty(\Omega)$. Then we can apply the Harnack inequality due to Trudinger in order to get $u > 0$ in Ω (cf. [7]). Finally, it follows from the result of Tolksdorf [24] that $u \in C^{1,\alpha}_{\text{loc}}(\Omega)$ (cf. [9]).

■

The case $\lambda = \lambda_1$

Let us consider the problem

(13.24)
$$\begin{aligned} -\Delta_p u &= \lambda_1 g(x)|u|^{p-2}u + f(x)|u|^{\gamma-2}u \text{ in } \Omega, \\ u &= 0 \text{ on } \partial\Omega. \end{aligned}$$

Keeping the notation from the previous section we consider the following conditional variational problem:

(P_λ) *Find maximizer $v_c \in W$ of the problem*

$$0 < M_{\lambda_1} = \sup\left\{\int_\Omega f(x)|v|^\gamma dx;\ H_{\lambda_1}(v) = 1\right\}.$$

We have $M_{\lambda_1} > 0$ due to $(f1)$. In this case, however, the set $W_{\lambda_1} = \{v \in W; H_{\lambda_1}(v) = 1\}$ is *unbounded* in W. So we are forced to require the following additional condition on f:

$(f2)$
$$\int_\Omega f(x)|u_1|^\gamma dx < 0$$

(here u_1 is the positive eigenfunction corresponding to the first eigenvalue $\lambda_1 > 0$ of $-\Delta_p$).

Proposition 13.2 *Assume* $(g0)$, $(f0)$, $(f1)$ *and* $(f2)$. *Then the problem* (P_{λ_1}) *has a nonnegative solution.*

Proof Let $\{v_n\}_{n=1}^\infty$ be the maximizing sequence of (P_{λ_1}), i.e.

(13.25)
$$H_{\lambda_1}(v_n) = 1,\ \int_\Omega f(x)|v_n|^\gamma dx = m_n \to M_{\lambda_1} > 0.$$

If $\{v_n\}_{n=1}^\infty$ is unbounded then we may assume that $\|v_n\|_W \to \infty$. Let us denote $t_n = \|v_n\|_W$ and set

$$v_n = t_n w_n \text{ with } \|w_n\|_W = 1.$$

So we get

$$H_{\lambda_1}(v_n) = t_n^p \left[\int_\Omega |\nabla w_n|^p dx - \lambda_1 \int_\Omega g(x)|w_n|^p dx \right] = 1$$

and due to (13.7)

$$0 \le \int_\Omega |\nabla w_n|^p dx - \lambda_1 \int_\Omega g(x)|w_n|^p dx = \frac{1}{t_n^p} \to 0. \tag{13.26}$$

In particular, we have

$$\liminf_{n\to\infty} \lambda_1 \int_\Omega g(x)|w_n|^p dx \ge 1. \tag{13.27}$$

We can assume that $w_n \rightharpoonup w_0$ in W for some $w_0 \in W$. Then (13.27) and $(g0)$ imply

$$\int_\Omega g(x)|w_0|^p dx \ge \frac{1}{\lambda_1}$$

and consequently $w_0 \not\equiv 0$. Since we have also

$$\int_\Omega |\nabla w_n|^p dx \le \liminf_{n\to\infty} \int_\Omega |\nabla w_n|^p dx = 1,$$

it follows from (13.26) that

$$0 \le \int_\Omega |\nabla w_n|^p dx - \lambda_1 \int_\Omega g(x)|w_0|^p dx \le 0.$$

Due to Lemma 13.1 w_0 is a multiple of the first eigenfunction u_1, i.e. there exists $k \ne 0$ such that

$$w_0(x) = ku_1(x).$$

On the other hand, thanks to (13.25) we get

$$\int_\Omega f(x)|v_n|^\gamma dx = t_n^\gamma \int_\Omega f(x)|w_n|^\gamma dx = m_n \to M_{\lambda_1} > 0,$$

i.e.

$$\int_\Omega f(x)|w_n|^\gamma dx = \frac{m_n}{t_n^\gamma} \to \ge 0$$

Due to $(f0)$ we have

$$\int_\Omega f(x)|w_0|^\gamma dx \ge 0$$

and consequently

$$\int_\Omega f(x)|u_1|^\gamma dx \ge 0$$

which contradicts $(f2)$. Hence the *maximizing sequence is bounded* and we may assume that

$$v_n \rightharpoonup v_c \text{ in } W.$$

By $(f0)$ we have

$$\int_\Omega f(x)|v_n|^\gamma dx \to \int_\Omega f(x)|v_c|^\gamma dx = M_{\lambda_1} > 0, \tag{13.28}$$

and hence $v_c \not\equiv 0$. It follows from (13.25), (13.7) and $(g0)$ that

$$0 \le \int_\Omega |\nabla v_c|^p dx - \lambda_1 \int_\Omega g(x)|v_c|^p dx \le 1.$$

We prove, first, that

$$0 < \int_\Omega |\nabla v_c|^p dx - \lambda_1 \int_\Omega g(x)|v_c|^p dx.$$

Indeed, otherwise the equality together with Lemma 13.1 would yield the existence of $k \ne 0$ such that $v_c(x) = ku_1(x)$. If we substitute it into (13.28), we get

$$|k|^\gamma \int_\Omega f(x)|u_1|^\gamma dx = M_{\lambda_1} > 0,$$

a contradiction with $(f2)$.

Now, we prove that

$$\int_\Omega |\nabla v_c|^p dx - \lambda_1 \int_\Omega g(x)|v_c|^p dx = 1. \tag{13.29}$$

Let us assume

$$0 < \int_\Omega |\nabla v_c|^p dx - \lambda_1 \int_\Omega g(x)|v_c|^p dx < 1.$$

Then there is $t_c > 1$ such that the function $\tilde{v}_c = t_c v_c$ satisfies $H_{\lambda_1}(\tilde{v}_c) = 1$ and simultaneously

$$\int_\Omega f(x)|\tilde{v}_n|^p dx = t_c^\gamma \int_\Omega f(x)|v_c|^p dx = t_c^\gamma . M_{\lambda_1} > M_{\lambda_1} = \sup\left\{\int_\Omega f(x)|v_n|^\gamma dx;\ H_{\lambda_1}(v) = 1\right\},$$

a contradiction. This proves (13.29) and hence v_c is the maximizer of (P_{λ_1}). We can assume that $v_c \ge 0$ in Ω due to Lemma 13.2.

■

Theorem 13.2 *Let $1 < p < \gamma < p^*$, $g, f \in L^\infty(\Omega)$ and $(f1)$, $(f2)$ be satisfied. Then the boundary value problem (13.24) has at least one positive weak solution $u \in W \cap L^\infty(\Omega)$. Moreover $u \in C^{1,\alpha}_{\mathrm{loc}}(\Omega)$.*

Proof of Theorem 13.2 is based on Proposition 13.6 and follows the same ideas as the proof of Theorem 13.1.

■

The case $\lambda > \lambda_1$

We consider the boundary value problem

(13.30) $$\begin{cases} -\Delta_p u = \lambda g(x)|u|^{p-2}u + f(x)|u|^{\gamma-2}u & \text{in } \Omega, \\ u = 0 & \text{on } \partial\Omega. \end{cases}$$

with $\lambda > \lambda_1$ but close enough to λ_1. The main result is formulated in the next theorem.

Theorem 13.3 *Let $1 < p < \gamma < p^*$, $g, f \in L^\infty(\Omega)$ and $(f1)$, $(f2)$ be satisfied. Then there exists $\varepsilon > 0$ such that for any $\lambda \in (\lambda_1, \lambda_1 + \varepsilon)$ the problem (13.30) has two distinct positive solutions $u_1, u_2 \in W \cap L^\infty(\Omega)$ and both belong to $C^{1,\alpha}_{\text{loc}}(\Omega)$.*

To prove this multiplicity result we consider two variational problems:

(P^1_λ) *Find maximizer $v_1 \in W$ of the problem*

$$M_\lambda = \sup\left\{\int_\Omega f(x)|v|^\gamma dx; \quad H_\lambda(v) = 1\right\}.$$

(P^2_λ) *Find minimizer $v_2 \in W$ of the problem*

$$m_\lambda = \inf\left\{H_\lambda(v); \quad \int_\Omega f(x)|v|^\gamma dx = -1\right\}.$$

Lemma 13.5 *Assume $(f1)$. The problem (P^1_λ) is equivalent to $(\tilde{P}^1_\lambda)$: Find maximizer $\tilde{v}_1 \in W$ of the problem*

$$0 < \tilde{M}_\lambda = \sup\left\{\int_\Omega f(x)|v|^\gamma dx; H_\lambda(v) \le 1\right\}.$$

Proof Note that $\tilde{M}_\lambda > 0$ due to the assumption $(f1)$. Let $\tilde{v}_1 \in W$ be the maximizer of $(\tilde{P}^1_\lambda)$ and

$$H_\lambda(\tilde{v}_1) < 1.$$

Then for arbitrary $w \in W$ there is sufficiently small $\eta > 0$ such that for $\tilde{v}_\eta = \tilde{v}_1 + \eta w$ we have

(13.31) $$H_\lambda(\tilde{v}_\eta) \le 1.$$

Then

(13.32) $$\int_\Omega f(x)|\tilde{v}_1 + \eta w|^\gamma dx = \\ = \int_\Omega f(x)|\tilde{v}_1|^\gamma dx + \gamma\eta\int_\Omega f(x)|\tilde{v}_1|^{\gamma-2}\tilde{v}_1 w\, dx + o(\eta) > \int_\Omega f(x)|\tilde{v}_1|^\gamma dx$$

if η is small enough and

$$\int_\Omega f(x)|\tilde{v}_1|^{\gamma-2}\tilde{v}_1 w dx > 0.$$

Note that this is valid if e.g. $w = \tilde{v}_1$ because we have

$$\int_\Omega f(x)|\tilde{v}_1|^{\gamma-2}\tilde{v}_1\tilde{v}_1 dx = \int_\Omega f(x)|\tilde{v}_1|^\gamma dx = \tilde{M}_\lambda > 0.$$

Hence (13.31), (13.32) are in contradiction with the fact that $\tilde{v}_1$ is maximizer of $(\tilde{P}^1_\lambda)$. ∎

Proposition 13.3 *Let $(g0)$, $(f0)$, $(f1)$ and $(f2)$ be satisfied. Then there exists $\varepsilon_1 > 0$ such that for any $\lambda \in (\lambda_1, \lambda_1+\varepsilon_1)$ the problem (P^1_λ) has a nonnegative solution $v_1 \in W$.*

Proof Due to Lemma 13.5 it suffices to show that there exists $\varepsilon_1 > 0$ such that $(\tilde{P}^1_\lambda)$ has a nonnegative solution for any $\lambda \in (\lambda_1, \lambda_1 + \varepsilon_1)$. We will prove this via contradiction. Assume that there is a sequence $\varepsilon_k \to 0$ such that for any $\lambda^k = \lambda_1 + \varepsilon_k$ the problem $(\tilde{P}^1_{\lambda_k})$ has no solution. Let us fix $k \in \mathbb{N}$ and consider the problem $(\tilde{P}^1_{\lambda^k})$. Let $\{v^k_n\}^\infty_{n=1}$ be a maximizing sequence of this problem, i.e.

$$\int_\Omega |\nabla v^k_n|^p dx - \lambda^k \int_\Omega g(x)|v^k_n|^p dx \le 1$$

and

$$\int_\Omega f(x)|v^k_n|^\gamma dx \to M_{\lambda^k} > 0$$

as $n \to \infty$. If the sequence $\{v^k_n\}^\infty_{n=1}$ is bounded, the variational problem $(\tilde{P}^1_{\lambda^k})$ has a (nonnegative) solution and this would be a contradiction. Indeed, if $\{v^k_n\}^\infty_{n=1}$ is bounded, we may assume

$$v^k_n \rightharpoonup v^k_0 \text{ in } W$$

as $n \to \infty$. Using $(g0)$, $(f0)$, $(f1)$ and repeating arguments from the proof of Proposition 13.6 we obtain

$$\int_\Omega f(x)|v^k_0|^\gamma dx = M_{\lambda^k} > 0,$$

$$\int_\Omega |\nabla v^k_0|^p dx - \lambda^k \int_\Omega g(x)|v^k_0|^p dx \le 1.$$

Hence v^k_0 is the solution of $(\tilde{P}^1_{\lambda^k})$.

Let us assume, now, that $\{v^k_n\}^\infty_{n=1}$ is unbounded in W. Then we may assume

$$\|v^k_n\|_W \to \infty \text{ as } n \to \infty.$$

Set $v^k_n = t^k_n w^k_n$ with $t^k_n = \|v^k_n\|_W$ and $\|w^k_n\|_W = 1$. We can assume again

$$w^k_n \rightharpoonup w^k_0 \text{ in } W.$$

We have

$$\int_\Omega f(x)|v^k_n|^\gamma dx = |t^k_n|^\gamma \int_\Omega f(x)|w^k_n|^\gamma dx \to M_k > 0$$

and therefore

(13.33) $$\int_\Omega f(x)|w_0^k|^\gamma dx \ge 0.$$

Futhermore

$$|t_n^k|^p \left[\int_\Omega |\nabla w_n^k|^p dx - \lambda^k \int_\Omega g(x)|w_n^k|^p dx\right] \le 1$$

and so

(13.34) $$\int_\Omega |\nabla w_n^k|^p dx - \lambda^k \int_\Omega g(x)|w_n^k|^p dx \le \frac{1}{|t_n^k|^p}.$$

This together with $(g0)$ and the weak lower semicontinuity of the norm in W imply

(13.35) $$\int_\Omega |\nabla w_0^k|^p dx - \lambda^k \int_\Omega g(x)|w_0^k|^p dx \le 0.$$

It follows from (13.34) that

$$\lambda^k \int_\Omega g(x)|w_n^k|^p dx \ge \int_\Omega |\nabla w_n^k|^p dx - \frac{1}{|t_n^k|^p}$$

and letting $n \to \infty$ we get from here (using $(g0)$ again):

(13.36) $$\lambda^k \int_\Omega g(x)|w_0^k|^p dx \ge 1.$$

Obviously, we have

(13.37) $$\int_\Omega |\nabla w_0^k|^p dx \le 1.$$

Now, we pass to the limit for $k \to \infty$. Then $\lambda^k \to \lambda_1$ and due to (13.37) we may assume that

$$w_0^k \rightharpoonup w_0 \text{ in } W$$

as $k \to \infty$. It follows from (13.36) and $(g0)$ that

(13.38) $$\lambda_1 \int_\Omega g(x)|w_0|^p dx \ge 1$$

and (13.35) yields

(13.39) $$\int_\Omega |\nabla w_0|^p dx - \lambda_1 \int_\Omega g(x)|w_0|^p dx \le 0.$$

The inequality (13.33) and $(f0)$ imply

(13.40) $$\int_\Omega f(x)|w_0|^\gamma dx \ge 0.$$

It follows from (13.38) that $w_0 \not\equiv 0$ and then (13.39) together with Lemma 13.1 imply

$$w_0(x) = ku_1(x), \quad k \ne 0.$$

Where $u_1 > 0$ is the eigenfunction corresponding to the first eigenvalue λ_1 of $-\Delta_p$. If we substitute this into (13.40), we get

$$|k|^\gamma \int_\Omega f(x)|u_1|^\gamma dx \geq 0$$

which contradicts $(f2)$.

Hence for some $\varepsilon_1 > 0$ the problem $(\tilde{P}^1_\lambda)$ (and also (P^1_λ) due to Lemma 13.5) has at least one (nonnegative) solution $v_1 \in W$ for any $\lambda \in (\lambda_1, \lambda_1 + \varepsilon_1)$.

■

Lemma 13.6 *Let $(f2)$ be satisfied. Then the set* $W^- = \left\{v \in W;\ \int_\Omega f(x)|v|^\gamma dx = -1\right\}$ *is nonempty and $m_\lambda < 0$ for any $\lambda > \lambda_1$.*

Proof For $v = tu_1$ (u_1 is the first eigenfunction of $-\Delta_p$) we get

$$\int_\Omega f(x)|v|^\gamma dx = |t|^\gamma \int_\Omega f(x)|u_1|^\gamma dx.$$

Due to $(f2)$ we have $\int_\Omega f(x)|u_1|^\gamma dx < 0$ and so we can find t_1, such that for $v_1 = t_1 u_1$:

$$\int_\Omega f(x)|v_1|^\gamma dx = |t_1|^\gamma \int_\Omega f(x)|u_1|^\gamma dx = -1.$$

We have also

$$H_\lambda(v_1) = |t_1|^p \left[\int_\Omega |\nabla u_1|^p dx - \lambda \int_\Omega g(x)|u_1|^p dx\right] = |t_1|^p(\lambda_1 - \lambda) \int_\Omega g(x)|u_1|^p dx < 0$$

for $\lambda > \lambda_1$ (see Lemma 13.1).

■

Proposition 13.4 *Let $(g0)$, $(f0)$, $(f1)$ and $(f2)$ be satisfied. then there exists $\varepsilon_2 > 0$ such that for any $\lambda \in (\lambda_1, \lambda_1 + \varepsilon_2)$ the problem (P^2_λ) has a nonnegative solution $v_2 \in W$.*

Proof Assume that the statement of Proposition 13.8 is not true. Then there exists a sequence $\varepsilon_k \to 0$ such that for $\lambda^k = \lambda_1 + \varepsilon_k$ the problem $(P^2_{\lambda^k})$ has no (nonnegative) solution. Let us consider a fixed $k \in \mathbb{N}$ and consider the minimizing sequence $\{v^k_n\}_{n=1}^\infty$ for $(P^2_{\lambda^k})$. Then

$$\int_\Omega f(x)|v^k_n|^\gamma dx = -1,$$

$$\int_\Omega |\nabla v^k_n|^p dx - \lambda^k \int_\Omega g(x)|v^k_n|^p dx \to m_{\lambda^k} < 0.$$

If $\{v^k_n\}_{n=1}^\infty$ is bounded then $m_{\lambda^k} > -\infty$ and similarly as in the proof of Proposition 13.7 we arrive at $v^k_n \rightharpoonup v^k_0$ as $n \to \infty$ and

$$\int_\Omega f(x)|v^k_0|^\gamma dx = -1, \quad \int_\Omega |\nabla v^k_0|^p dx - \lambda^k \int_\Omega g(x)|v^k_0|^p dx = m_{\lambda^k}.$$

Hence v_0^k is the solution of $(P^2_{\lambda^k})$, a contradiction.

Let us assume, now, that $\{v_n^k\}_{n=1}^{\infty}$ is unbounded. Similarly as in the proof of Proposition 6.1 we arrive at $v_n^k = t_n^k w_n^k$, $t_n^k = \|v_n^k\|_W \to \infty$, $\|w_n^k\|_W = 1$, $w_n^k \rightharpoonup v_0^k$ in W as $n \to \infty$. In this case we have

$$-1 = \int_\Omega f(x)|v_n^k|^\gamma dx = |t_n^k|^\gamma \int_\Omega f(x)|w_n^k|^\gamma dx,$$

i.e.

$$\int_\Omega f(x)|w_n^k|^\gamma dx = -\frac{1}{|t_n^k|^\gamma} \to 0$$

as $n \to \infty$. Due to $(f0)$, we have

$$\int_\Omega f(x)|w_0^k|^\gamma dx = 0. \tag{13.41}$$

On the other hand,

$$H_{\lambda^k}(v_n^k) = (t_n^k)^k \left[\int_\Omega |\nabla w_n^k|^p dx - \lambda^k \int_\Omega g(x)|w_n^k|^p dx\right] \le 0$$

and concequently

$$\int_\Omega |\nabla w_n^k|^p dx - \lambda^k \int_\Omega g(x)|w_n^k|^p dx \le 0. \tag{13.42}$$

Simillarly as in the proof of Proposition 13.7 we get from (13.42) that

$$\int_\Omega f(x)|\nabla w_0^k|^p dx - \lambda^k \int_\Omega g(x)|w_0^k|^p dx \le 0 \tag{13.43}$$

and

$$\lambda^k \int_\Omega g(x)|w_0^k|^p dx \ge 1. \tag{13.44}$$

Now, we shall pass to the limit for $k \to \infty$, i.e. we have $\lambda^k \to \lambda_1$. We may assume again that

$$w_0^k \rightharpoonup w_0 \text{ in } W$$

as $k \to \infty$. Using $(f0)$, $(g0)$ we get from (13.42), (13.43) and (13.44) that

$$\lambda_1 \int_\Omega g(x)|w_0|^p dx \ge 1 \quad (\text{i.e. } w_0 \not\equiv 0), \tag{13.45}$$

$$\begin{aligned} &\int_\Omega |\nabla w_0|^p dx - \lambda_1 \int_\Omega g(x)|w_0|^p dx \le \\ &\le \liminf_{k\to\infty} \left[\int_\Omega |\nabla w_0^k|^p dx - \lambda_1 \int_\Omega g(x)|w_0^k|^p dx\right] \le 0, \end{aligned} \tag{13.46}$$

and

$$\int_\Omega f(x)|w_0^k|^\gamma dx = \lim_{k\to\infty} \int_\Omega f(x)|w_0^k|^\gamma dx = 0. \tag{13.47}$$

Now (13.45), (13.46) imply $w_0 = ku_1$, with $k \neq 0$. If we substitute w_0 into (13.47) we get

$$\int_\Omega f(x)|u_1|^\gamma dx = 0,$$

a contradiction with $(f2)$.

It follows from above consedirations that there exists $\varepsilon_2 > 0$ such that the problem (P_λ^2) has at least one (nonnegative) solution $v_2 \in W$ for any $\lambda \in (\lambda_1, \lambda_1 + \varepsilon_2)$.

■

Proof of Theorem 13.3 Set $\varepsilon = \min\{\varepsilon_1, \varepsilon_2\}$ and consider $\lambda \in (\lambda_1, \lambda_1 + \varepsilon)$. Applying the fibering method we get weak solutions of (13.30) as in previous sections. Namely, let $r_i > 0$ be determined by (13.12) for $v := v_i$, $i = 1, 2$. Then

$$u_i = r_i v_i, \quad i = 1, 2$$

are nonnegative weak solutions to (13.30). The first weak solution u_1 belongs to the set

$$W_1 = \left\{ u \in W;\ \int_\Omega |\nabla u|^p dx - \lambda \int_\Omega g(x)|u|^p dx > 0 \right\}$$

because

$$\int_\Omega |\nabla u_1|^p dx - \lambda \int_\Omega g(x)|u_1|^p dx = (r_1)^p H_\lambda(v_1) = (r_1)^p > 0.$$

The second weak solution u_2 belongs to the set

$$W_2 = \left\{ u \in W;\ \int_\Omega |\nabla u|^p dx - \lambda \int_\Omega g(x)|u|^p dx \leq 0 \right\}$$

because

$$\int_\Omega |\nabla u_2|^p dx - \lambda \int_\Omega g(x)|u_2|^p dx = (r_2)^p H_\lambda(v_2) \leq 0.$$

Since

$$W_1 \cap W_2 = \emptyset,$$

we have

$$u_1 \not\equiv u_2.$$

Other properties of u_i, $i = 1, 2$ (i.e. positivity, $L^\infty-$bound and $C^{1,\alpha}_{\text{loc}}-$regularity) follow in the same way as above.

■

14 Existence of positive solutions of nonlinear Neumann problem

Now we consider the applications of the fibering method to the nonlinear Neumann problem.

Here we follow the paper by A. Tesei and S. Pohozaev [23], where this problem is considered in more general setting. In its turn this paper is generalizing some results from [5].

Let us consider the BVP

$$(14.1)\qquad \begin{cases} \Delta u + (\nabla\Phi, \nabla u) + a(x)|u|^{p-2}u = 0, u \geq 0, & \text{in } \Omega, \\ \dfrac{\partial u}{\partial \nu} = 0 & \text{on } \partial\Omega \end{cases}$$

under the following assumptions:
($A1$) $a \in L^{\infty}(\Omega)$,
($A2$) $a^{+}(x) = \max(a(x), 0) \not\equiv 0$,
($A3$) $a(x)\rho(x) < 0$, with $\rho(x) = e^{\psi(x)}$,
with $2 < p < 2^*$, $\psi \in C^1(\overline{\Omega})$.

Theorem 14.1 *Let the assumptions* ($A1$) – ($A3$) *and* $2 < p < 2^*$ *be satisfied. Then there exists a nonnegative (nontrivial) solution to (14.1).*

We prove this theorems, by using of the fibering method. The proof is divided into some steps.

I–st step

We consider the functional E defined by

$$E(u) = -\frac{1}{2}\int_{\Omega} |\nabla u|^2 \rho\, dx + \frac{1}{p}\int_{\Omega} a(x)|u|^p \rho\, dx$$

for $u \in W_2^1(\Omega)$. Thanks to ($A1$) and $2 < p < 2^*$ this Euler functional is well posed on the Sobolev space $W_2^1(\Omega)$. Following the fibering method we set

$$u(t) = rv(x), r \in \mathbb{R}, v \in W_2^1(\Omega).$$

Then $E(u)$ takes the form

$$E(v) = -\frac{1}{2}r^2 \int_{\Omega} |\nabla v|^2 \rho dx + \frac{|r|^p}{p}\int_{\Omega} a(x)|v|^p \rho\, dx.$$

Now we introduce the constraint

$$(14.1^*)\qquad \int_{\Omega} |\nabla v|^2 \rho\, dx = 1.$$

This functional under the condition (14.1*) reduces to

$$(14.2)\qquad \tilde{E} = -\frac{1}{2}r^2 + \frac{1}{p}|r|^p \int_{\Omega} a(x)|v|^p \rho dx.$$

The bifurcation equation

$$\frac{d\tilde{E}}{dr} \equiv \tilde{E}'_r = 0$$

becomes

$$-r + |r|^{p-2}r \int_\Omega a(x)|u|^p \rho dx = 0.$$

From here we derive for $r \neq 0$

$$r(v) = \pm \frac{1}{[\tilde{E}_0(v)]^{\frac{1}{p-2}}} \tag{14.3}$$

with

$$\tilde{E}_0(v) := \int_\Omega a(x)|v|^p \rho\, dx. \tag{14.4}$$

Then $\tilde{E}$ takes the form

$$\tilde{E}(v) = \left(\frac{1}{p} - \frac{1}{2}\right) \left[\tilde{E}_0(v)\right]^{-\frac{2}{p-2}}. \tag{14.5}$$

II-nd step

Now we consider the extremal point of $\tilde{E}_0(v)$ under (14.1*).

Namely

$$\{\tilde{E}_0(v) | v \in W_2^1(\Omega), \int_\Omega |\nabla v|^2 \rho^r\, dx = 1\} = M_0. \tag{14.6}$$

Lemma 14.1 *The following inequality holds*

$$M_0 > 0.$$

The proof follows from $(A2)$ immediately.

Lemma 14.2 *The following inequality holds*

$$M_0 < +\infty$$

and the maximizing sequence of (14.6) is bounded.

Proof It sufficies to prove the boundness of the maximizing sequence. Let

$$\int_\Omega a(x)|v_r|^p \rho(x)\, dx \to M_0 \tag{14.7}$$

and

$$\int_\Omega |\nabla v_n|^p \rho\, dx = 1.$$

Put

(14.8)
$$v_n(x) = C_n + \overline{v}_n(x)$$

where C_n are constants and

(14.9)
$$\int_\Omega \overline{v}_n \rho \, dx = 0.$$

Then by virtue of the Sobolev imbedding theorems (the Poincare inequality) we have

(14.10)
$$\left| \int_\Omega a(x) |\overline{v}_n|^p \rho \, dx \right| \leq M_1$$

thanks to (14.8) and (14.9). Here M does not depend on $\overline{v}_n$. Now we suppose to the contrary that

$$\|v_n\|_{W^{1,2}(\Omega)} \to +\infty.$$

Due to (14.8) and (14.10) it follows that

(14.11)
$$C_n \to \infty.$$

Further, we have

$$\int_\Omega a(x)|v_n(x)|^p \rho(x) \, dx = \int_\Omega a(x)|C_n + \overline{v}_n(x)|^p \rho(x) \, dx = |C_n|^p \int_\Omega a(x)|1 + \frac{\overline{v}_n(x)}{C_n}|^p \rho(x) \, dx.$$

Then by (14.10), (14.11) and (14.3) we get

$$\int_\Omega a(x)|1 + \frac{\overline{v}_n(x)}{C_n}|^p \rho(x) \, dx \to \int_\Omega a(x)\rho(x) \, dx < 0.$$

Thus the limit (14.7) satisfies

$$M_0 < 0.$$

This is a contradiction with Lemma 14.1. So we proved that the maximizing sequence $\{v_n\}$ is bounded in $W^{1,2}(\Omega)$.

■

Lemma 14.3 *There exists a maximizer of (14.6).*

Proof By $2 < p < 2^*$ and $(A1)$ we can use the following Kondrashev imbedding theorem due to compact imbedding $W^{1,2}(\Omega) \subset\subset L^p_\rho$.

So, we proved the existence of $v_0 \in W^{1,2}(\Omega)$ such that

$$\sup\{\tilde{E}_0(v) | v \in W^{1,2}(\Omega), \int_\Omega |\nabla v|^2 \, dx \leq 1\} = \tilde{E}_0(v_0)$$

with

$$\int_\Omega |\nabla v_0|^2 \rho \, dx < 1.$$

Now we prove that $\int_\Omega |\nabla v_0|^2 \rho \, dx = 1$. Indeed if

$$\int_\Omega |\nabla v_0|^2 \rho \, dx < 1$$

we consider

$$\tilde{v}_0 = t v_0$$

with $t > 1$ such that

$$\int_\Omega |\nabla \tilde{v}_0|^2 \rho \, dx = |t|^2 \int_\Omega |\nabla v_0|^2 \rho \, dx = 1.$$

We note that

$$0 < \int_\Omega |\nabla v_0|^2 \rho \, dx.$$

Because otherwise, i.e. $\int_\Omega |\nabla v_0|^2 \rho \, dx = 0$, we would have

$$v_0(x) = C_0$$

and

$$\int_\Omega a(x)|v_0|^p \rho \, dx = |C_0|^p \int_\Omega a(x)\rho(x) \, dx < 0$$

by $(A3)$. But this is impossible by Lemma 14.1. Then for a new element $\tilde{v}_0$ we find

$$\int_\Omega a(x)|\tilde{v}_0|^p \rho(x) dx = |t|^p \int_\Omega a(x)|v_0|^p \rho(x) \, dx = |t|^p M_0 > M_0.$$

This is a contradiction with the definition of M_0 by (14.6). Thus we proved the existence of a maximizer v_0 of (14.6) with

$$\tilde{E}_0(v_0) > 0.$$

■

III-rd step

Further from the general properties of the Sobolev space $W^{1,2}(\Omega)$ we have $|v_0| \in W^{1,2}(\Omega)$ and

$$\int_\Omega |D|v_0||^2 \rho \, dx = \int_\Omega |Dv_0|^2 \rho \, dx.$$

Thus we can assume that the maximizer is always nonnegative: $v_0 \geq 0$ in Ω. Now, from the fibering method it follows that

$$u_0(x) = r_0 v_0(x) \geq 0$$

with

$$r_0 = + \left[\tilde{E}_0(v)\right]^{-\frac{1}{p-2}}$$

is a nonnegative solution to the BVP (14.1). We note that $u_0 \not\equiv 0$ because

$$\int_\Omega a|u_0|^p \rho \, dx = |r_0|^p \int_\Omega a|v_0|^p \rho \, dx = [\tilde{E}_0(v_0)]^{\frac{2}{p-2}} > 0.$$

The proof of Theorem 14.1 is completed.

The case $f = a(x)|u|^{p-2}$ with $1 < p < 2$

Here we consider the BVP (14.1) with $1 < p < 2$. In this case we have for the Euler functional $E(u)$ the same representation (14.2) but with $1 < p < 2$. Now the graph of $\tilde{E}$ in r looks as in fig. 6.

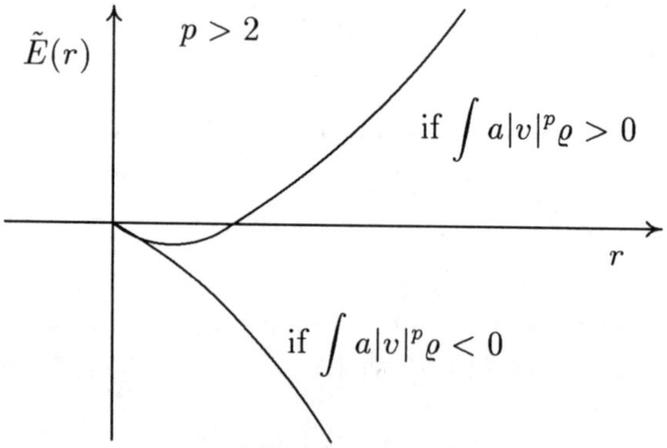

Figure 5:

Theorem 14.2 *Let the assumptions (A1) – (A3) be satisfied and $1 < p < 2$. Then there exists a nonnegative (nontrivial) solution to (14.1).*

The proof coincides with the proof of Theorem 14.1.

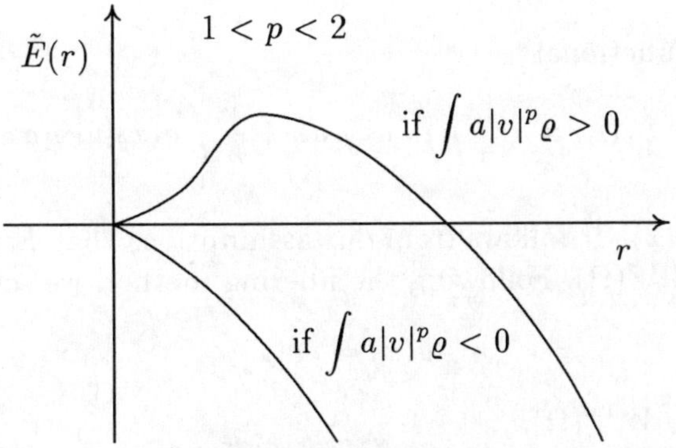

Figure 6:

Existence of positive solutions of nonlinear Neumann problem for the $p-$Laplacian

Here we study the existence at positive solutions of the BVP

$$\text{(14.12)} \qquad \begin{cases} \Delta_p u + \langle \nabla \Phi, \nabla u \rangle |\nabla u|^{p-2} + a(x)|u|^{q-2}u = 0 & \text{in bdd. } \Omega \subset \tilde{\mathbb{R}}, \\ \dfrac{\partial u}{\partial v} = 0. \end{cases}$$

Here $\Delta_p u := \operatorname{div}(|\nabla u|^{p-2}\nabla u)$ with $p > 1$ and

$$\Phi \in C^1(\overline{\Omega}).$$

We consider this problem under the following assumptions:

$(A_p 1)$ $\quad a \in L^\infty(\Omega)$,

$(A_p 2)$ $\quad a_+(x) \not\equiv 0$,

$(A_p 3)$ $\quad \int\limits_\Omega a(x)\rho(x)\,dx < 0$,

where $\rho(x) = e^{\phi(x)}$.

Concerning q we suppose

$$\text{(14.13)} \qquad 1 < q < p^*, \quad q \neq p, \quad p^* = \begin{cases} \dfrac{Np}{N-p}, & \text{for } N > p, \\ +\infty, & \text{for } N = p. \end{cases}$$

Theorem 14.3 *Let the assumptions* $(A_p 1) - (A_p 3)$ *and (14.13) be satisfied. Then the BVP (14.12) has a nonnegative (nontrivial) solution.*

$$u \in W^{1,p}(\Omega) \cap L^\infty(\Omega) \cap C^{1,\alpha}_{\text{loc}}(\Omega).$$

The proof is based on the variational approach and it is divided into five steps.

I-st step

We have the Euler functional

$$E_p(u) = -\frac{1}{p}\int_\Omega |\nabla u|^p \rho\, dx + \frac{1}{q}\int_\Omega a(x)|u|^q \rho\, dx$$

associated with (14.12). It follows from our assumptions that $E_p(u)$ is well defined on the Sobolev space $W^{1,p}(\Omega)$. Following the fibering method we set

$$u(x) = rv(x), \tag{14.14}$$

where $r \in \mathbb{R}$ and $v \in W^{1,p}(\Omega)$.

Then we get

$$E_p = -\frac{|r|^p}{p}\int_\Omega |\nabla v|^p \rho\, dx + \frac{|r|^q}{q}\int_\Omega a(x)|v|^q \rho\, dx.$$

Now we introduce the constraint

$$\int_\Omega |\nabla v|^p \rho\, dx = 1. \tag{14.15}$$

The functional E_p under (14.15) takes the form

$$E_p = -\frac{|r|^p}{p} + \frac{|r|^q}{q}\int_\Omega a(x)|v|^q \rho\, dx. \tag{14.16}$$

The bifurcation equation is the following one

$$\frac{dE_p}{dr} = -|r|^{p-2}r + |r|^{q-2}r\int_\Omega a(x)|v|^q \rho\, dx = 0;$$

for $r \neq 0$ we find

$$r = \pm\frac{1}{\left(\int_\Omega a(x)|v|^q \rho\, dx\right)^{-\frac{1}{p-q}}} = \pm\left(\int_\Omega a(x)|v|^q \rho\, dx\right)^{\frac{1}{p-q}} \tag{14.17}$$

for

$$\int_\Omega a(x)|v|^q \rho\, dx > 0.$$

Now, substitute (14.17) into (14.16) we find

$$\tilde{E}(v) = \left(\frac{1}{q} - \frac{1}{p}\right)[E_1(v)]^{+\frac{p}{p-q}}, \quad E_1(v) \equiv \int_\Omega a(x)|v|^q \rho\, dx. \tag{14.18}$$

II-nd step

Let us consider the variational problem

$$\sup\{E_1(v) | \int_\Omega |\nabla v|^p \rho \, dx \le 1\} \stackrel{\text{def}}{=} M_1. \tag{14.19}$$

Lemma 14.4 *There exists an extreme point $v_0 \in W_2^1$ with $\int_\Omega |\nabla v_0|^p \rho \, dx = 1$ of the variational problem (14.19).*

Proof Thanks to $(A_p 2)$ we have

$$M_1 > 0. \tag{14.20}$$

Let $\{u_n\}$ be a maximizing sequence of (14.19). Set

$$v_n(x) = C_n + \overline{v}_n(x), \tag{14.21}$$

where

$$\int_\Omega \overline{v}_n(x) \rho \, dx = 0. \tag{14.22}$$

By

$$\int_\Omega |\nabla v_n|^p \rho \, dx = \int_\Omega |\nabla \overline{v}_n|^p \rho \, dx \le 1 \tag{14.23}$$

and (14.22) we have

$$E_1(\overline{v}_n) = \int_\Omega a(x) |\overline{v}_n|^q \rho \, dx \le C < \infty \tag{14.24}$$

due to the Sobolev imbedding theorem (the Poincare inequality), where C–doesn't depend on $\overline{v}_n$.

Now we prove that the maximizing sequence $\{u_n\}$ is bounded in $W^{1,p}$.

Suppose the contrary. Then by (14.23) and (14.21) we get

$$C_n \to \infty. \tag{14.25}$$

Substitute (14.21) into E_1 we find

$$\begin{aligned} E_1(v_n) &= \int_\Omega a(x) |C_n + \overline{v}_n(x)|^q \rho \, dx = \\ &= |C_n|^p \int_\Omega a(x) |1 + \frac{\overline{v}_n(x)}{C_n}|^q \rho \, dx. \end{aligned} \tag{14.26}$$

From (14.24) by (14.25) and $(A3)$ it follows that

$$\int_\Omega a(x) |1 + \frac{\overline{v}_n(x)}{C_n}|^q \rho \, dx \to \int_\Omega a(x) \rho(x) \, dx < 0.$$

Then (14.26) implies that

$$E_0(v_u) \to -\infty.$$

This contradicts (14.20). Now from the boundness of $\{v_n\}$ and reflexivity of $W^{1,2}(\Omega)$ and week continuity of $E_1(v)$ we obtain the existence of $v_0 \in W^{1,p}(\Omega)$ such that

$$\sup\{E_1(v) | \int_\Omega |\nabla v_0|^p \rho\, dx \leq 1\} = E_1(v_0) > 0.$$

Now, by standard arguments we prove that

$$\int_\Omega |\nabla v_0|^p \rho\, dx = 1$$

and

$$v_0 \geq 0 \text{ in } \Omega.$$

■

III-rd step

By $E_1(v_0) > 0$ it follows from (14.18) that v_0 is an extremal point of $\tilde{E}(v)$. Then by virtue of the fibering method we derive that

$$u_0(x) = r_0 v_0(x)$$

with

$$r_0 = +E_1(v)^{\frac{1}{p-q}}$$

as a nonnegative solution (14.12).

IV-th step

For $u_0 \geq 0$ we have

$$\int_\Omega a(x)|u_0|^p \rho\, dx = |r_0|^q \int_\Omega a(x)|v_0|^q \rho\, dx =$$

$$= [E_1(v_0)]^{-\frac{q}{p-q}} E_1(v_0) = [E_1(v_0)]^{\frac{p}{p-q}} > 0.$$

Thus the solution $u_0 \geq 0$ is a nontrivial one.

V-th step

We obtain nonnegative nontrivial solution u_0 of (14.12) from the Sobolev space $W^{1,p}(\Omega)$, thus $u_0 \geq 0$ is a week solution of (14.12). Now following the bootstrap argument (used e.g. in [9]) we prove $u \in L^\infty(\Omega)$. Then from the result of Tolksdorf [24] it follows that $u \in C^{1,\alpha}_{\text{loc}}(\Omega)$.
Theorem 14.3 is proved.

15 A test for the absence of a solution. Example

We first present a scheme for getting sufficient conditions for the absence of a solution in the case of nonlinear variational problems.

Let us consider the variational problem from Section 8. On a Banach space X with norm differentiable away from the origin a functional is equivalent to the existence of a solution (t, v) with $t \in \mathbb{R} \setminus \{0\}$ and $v \in S$ of the following system of equations:

$$\begin{cases} p|t|^{p-1} \operatorname{sgn} t + \langle f_1'(tv), v \rangle = 0, & t \neq 0, \\ |t|^{p-1} \operatorname{sgn} t(\|v\|^p)' + f_1'(tv) = 0, & v \in S. \end{cases}$$

Suppose now that w is some element of X. Then from this system we get the following system of two scalar equations:

$$\begin{cases} p|t|^{p-1} \operatorname{sgn} t + \langle f_1'(tv), v \rangle = 0, \\ |t|^{p-1} \operatorname{sgn} t\langle (\|v\|^p)', w \rangle + \langle f_1'(tv), w \rangle = 0. \end{cases} \tag{15.1}$$

This gives us the following test for the absence of a nonzero solution in X of the equation

$$f_0'(u) + f_1'(u) = 0. \tag{15.2}$$

Theorem 15.1 *Let f_0 and f_1 be the functionals defined above, and suppose that there exists an element $w \in X$ such that system (15.1) is inconsistent whatever the values of $t \in \mathbb{R} \setminus \{0\}$ and $v \in S$ are. Then the equation (15.2) does not have nontrivial solutions in X.*

Remark The zero solution of (15.3) obviously does not exist if

$$f_0'(0) + f_1'(0) \neq 0.$$

As an application of this test we again consider the boundary value problem with quadratic nonlinearity in Section 11. System (15.1) for the equivalent problem (11.2) takes the form

$$\begin{cases} t - \displaystyle\int_\Omega (tv + H)^2 v \, dx + \int_\Omega \varphi v \, dx = 0, \\ -t \displaystyle\int_\Omega v \Delta\psi \, dx - \int_\Omega (tv + H)^2 \psi \, dx + \int_\Omega \varphi\psi \, dx = 0. \end{cases}$$

Here ψ is a function in $\overset{\circ}{W}{}_2^1(\Omega)$ that is arbitrary for a while. The first equation in this system clearly can be obtained from the second one by setting $\psi = v$.

Therefore, without fixing for the time being the function ψ in $\overset{\circ}{W}{}_2^1(\Omega)$, we consider the second scalar equation with respect to t. This equation clearly does not have a real solution if there exists a function $\psi \in \overset{\circ}{W}{}_2^1(\Omega)$ such that

$$\left(\int_\Omega (\Delta\psi + 2H\psi) v \, dx \right)^2 < 4 \int_\Omega (H^2 - \varphi)\psi \, dx \int_\Omega \psi v^2 \, dx, \quad \forall v \in S. \tag{15.3}$$

On the other hand, if $\psi(x) \geq 0$ in Ω, then

$$\left(\int_\Omega (\Delta\psi + 2H\psi)v\ dx\right)^2 = \left(\int_\Omega \frac{\Delta\psi + 2H\psi}{\sqrt{\psi}} \sqrt{\psi} v\ dx\right)^2 \leq$$

$$\leq \int_\Omega \frac{(\Delta\psi + 2H\psi)^2}{\psi} dx \int_\Omega \psi v^2\ dx.$$

Hence (15.3) holds if there exists a function $\psi \geq 0$ in $\overset{\circ}{W}{}^1_2(\Omega)$ such that

$$\int_\Omega \frac{(\Delta\psi + 2H\psi)^2}{\psi} dx < 4\int_\Omega (H^2 - \varphi)\psi\ dx.$$

Accordingly, we get the following test for the absence of a real solution $\Phi \in \overset{\circ}{W}{}^1_2(\Omega)$ of the problem (11.1). Suppose that there exists a function $\varphi \geq 0$ in $\overset{\circ}{W}{}^1_2(\Omega)$ such that

$$\text{(15.4)} \qquad \int_\Omega \left[\frac{(\Delta\psi)^2}{\psi} + 4H\Delta\psi + 4\varphi\psi\right] dx < 0.$$

Then the boundary value problem (11.1) does not have a real solution in $\overset{\circ}{W}{}^1_2(\Omega)$. Choosing now corresponding functions $\psi \geq 0$ in $\overset{\circ}{W}{}^1_2(\Omega)$, we get concrete tests for the absence of a real solution of the boundary value problem (11.1) in $W_2^1(\Omega)$.

We remark that the test (15.4), in contrast to the traditional tests for quasilinear elliptic equations of second order, is not a pointwise test but an integral test. We explain this using the following special example.

Consider the boundary value problem (10.1), where Ω_1 is the unit disk around the origin in the plane $\mathbb{R}^2$, $\varphi(x) \equiv 0$, and the boundary function h is equal to $A\cos\theta$ in polar coordinates, i.e.,

$$\text{(15.5)} \qquad \begin{cases} \Delta\Phi + \Phi^2 = 0 & \text{in } \Omega_1 = \{(x,y) \in \mathbb{R}^2 | r^2 = x^2 + y^2 < 1\}, \\ \Phi = A\cos\theta & \text{on } \partial\Omega_1. \end{cases}$$

Here A is an arbitrary parameter. The choice of this particular example is due to two circumstances. First, this problem is given without analysis in a number of books. Second – and this is the main thing – the traditional tests for the absence of a real solution are not applicable to problem (15.5), since the mean of the boundary values is equal to zero:

$$\int_0^{2\pi} A\cos\theta\ d\theta = 0,$$

so that here the boundary values do not have definite sign.

For problem (15.5) the inequality (15.4) takes the form

$$\text{(15.6)} \qquad \int_{\theta=0}^{2\pi}\int_{r=0}^{1} \left[\frac{(\Delta\psi)^2}{\psi} + 4Ar\cos\theta.\Delta\psi\right] r\ dr\ d\theta < 0.$$

We now choose ψ to be a solution of the following problem with parameter $\tau > 0$:

$$\Delta\psi = -(\tau + r\cos\theta)(1 - r^2) \text{ in } \Omega_1, \quad \psi = 0 \text{ on } \partial\Omega_1.$$

This solution can be written explicitly, and $\psi \geq 0$ for $\tau \geq \frac{1}{3}$. We substitute this function ψ (which depends on $\tau \geq \frac{1}{3}$) into (15.6). Then we get a parametric inequality for A with $\tau \geq \frac{1}{3}$, and it yields the following estimate for $|A|$ when $\tau = \frac{(1+\sqrt{\frac{5}{2}})}{3}$:

$$|A| > 20.65.$$

If A satisfies this inequality, then the boundary value problem (15.5) does not have a real solution in $W_2^1(\Omega_1)$.

16 Conclusion

We showed the applications of the fibering method to

1. the existence problem,
2. the existence of multiple solutions,
3. the existence of infinitely many solutions,
4. the problem nonexistence of the solution.

The above mentioned results have been obtained by using of the one–parameter fibering method. If we consider the imbedding of the original space X in to $\tilde{X} = \mathbb{R}^k \times X$ we get the $k-$parametric fibering method. As an application of 2–parametric fibering method we consider the following nonlinear problem.

The problem with a nonlinear resonance. The authors of [7] considered a semi-linear elliptic problem at resonance that is a linear problem with spectral value of the parameter and with a nonlinear perturbation "small" at infinity. We consider the problem when the perturbation is superlinear and anticoercive. It is natural to base the investigation of such problems no longer on a one–parameter fibration but on a multiparameter one. The simplest variant of a two–parameter fibration is used in this problem.

Let Ω be a bounded region in $\mathbb{R}^n$, $n \leq 3$, with locally Lipschitz boundary $\partial\Omega$. We consider in $\overset{\circ}{W}{}_2^1(\Omega)$ the following boundary value problem

$$\Delta u + \lambda_1 u + u^3 = 0 \text{ in } \Omega, \quad u = 0 \text{ on } \partial\Omega; \tag{16.1}$$

here λ_1 is the first eigenvalue of the problem

$$\Delta e_1 + \lambda_1 e_1 = 0 \text{ in } \Omega, \quad e_1 = 0 \text{ on } \partial\Omega$$

with eigenfunction $e_1 \in \overset{\circ}{W}{}_2^1(\Omega)$, $\int_\Omega |\nabla e_1|^2 dx = \lambda_1$.

We write $u = te_1 + u^1$, where $t \in \mathbb{R}$, $\langle e_1, u^1 \rangle = 0$, and we set $u^1 = sv$ with $s \in \mathbb{R}$. Then the functional f corresponding to the problem (16.1) takes the form

$$f(t,s,v) = \frac{1}{2}s^2 \left(\int_\Omega |\nabla v|^2 dx - \lambda_1 \int_\Omega v^2 \, dx \right) - \frac{1}{4} \int_\Omega (te_1 + sv)^4 \, dx;$$

associated functional $\tilde{f}$ takes the form

$$\tilde{f}(v) = \max_{s>0, t\in\mathbb{R}} f(t,s,v),$$

which belongs to the class C^2 for $v \neq 0$ and $\langle e_1, v \rangle = 0$. Here the fibering functional H is defined by

$$H(v) \equiv f_{ss}(t,s,v) = -1, \tag{16.2}$$

where $t = t(v)$ and $s = s(v) > 0$ are uniquely determined from the relation

$$f(t(v), s(v), v) = \max_{s>0, t\in\mathbb{R}} f(t,s,v)$$

with respect to $v \neq 0$ and for a fixed eigenfunction e_1, $\langle e_1, v \rangle = 0$.

We apply the Lyusternik–Schnirelmann method to the even functional $\tilde{f}$, considered on the manifold (16.2) for an even functional H. Then we get that the boundary value problem (16.1) has a countable set of nontrivial solutions.

Remark The application at the fibering method to the system of nonlinear ordinary differential equations can be find in [20]. Some other applications of the fibering method can be found in [4] and [17].

Acknowledgement

The author wants to express his gratitude to Professor Pavel Drábek for his help in the preparation of this material for the print.

References

[1] R. A. Adams, *Sobolev Spaces,* Academia Press, N. Y. 1975.

[2] A. Ambrosetti, R. H. Rabinowitz, *Dual variational methods in critical point theory and applications,* J. Functional Anal. **14** (1973), 349–381.

[3] A. Anane, *Simplicité et isolation de la premiere valeur propre du p−laplacien avec poids,* C. R. Acad. Sci. Paris 305. Sér I (1987), 725–728.

[4] A. Bahri, H. Berestycki, *Points critiques de perturbations de fonctionnelles paires et applications,* C. R. Acad. Sci. Paris Sér. A–B **291** (1980), A189–A192.

[5] C. Bandle, M. A. Pozio, A. Tesei, *Existence and Uniqueness of Solutions of Nonlinear Neumann Problems,* Math. Z., **199,** (1988), 257–278.

[6] G. Barles, *Remarks on uniqueness results of the first eigenvalue of the p–Laplacian,* Ann. Fac. Sci. Toulouse 9, No. 1 (1988), 65–75.

[7] P. Bartolo, V. Benci, D. Fortunato, *Abstract critical point theorems and application, to some problelms with strony resonance at infinity,* Nonlinear Analysis, T.M.A. 7, (1983), 981-1012.

[8] P. Drábek, *Solvability and Bifurcations of Nonlinear Equations,* Pitman Research Notes in Math. 264, Longman, Harlow 1992.

[9] P. Drábek, A. Kufner, F. Nicolosi, *Quasilinear Elliptic Equations with Degenerations and Singularities,* Walter de Gruyter Series in Nonlinear Analysis and Applications, Vol. 5, New York - Berlin 1997.

[10] P. Drábek, Y. Huang, *Bigurcation problems for the p–Laplacian in* $\mathbb{R}^N$, to appear in Trans. Amer. Math. Society.

[11] P. Drábek, Y. Huang, *Multiple positive solutions of quasilinear elliptic equations in* $\mathbb{R}^N$, to appear in Nonlinear Analysis, Theory, Methods & Applications.

[12] P. Drábek, S. I. Pohozaev, *Positive solutions for the p–Laplacian: application of the fibering method,* to appear in Proc. Royal Soc. of Edinburg.

[13] S. Fučík, A. Kufner, *Nonlinear Differential Equations,* Elsevier, Holland 1980.

[14] P. Lindqvist, *On the equation* $\mathrm{div}(|\nabla u|^{p-2}\nabla u) + \lambda |u|^{p-2}u = 0$, Proc. Amer. Math. Society 109, No. 1 (1990), 157–164.

[15] J. L. Lions, *Quelques méthodes de résolution des problemes aux limites non linéarires,* Dunod and Gauthier–Villars, Paris, 1969.

[16] L. Nirenberg, *Topics in nonlinear functional analysis,* Lecture Notes, Courant Inst. Math. Sci., New York Univ., New York, 1974.

[17] P. I. Plotnikov, *The existence of a countable set of periodic solutions of the problem of forced oscillations for a weakly nonlinear wave equation,* Mat. Sb. **136 (178)** (1988), 546-560; English transl. in Math. USSR Sb. **64** (1989).

[18] S. I. Pohozaev, *On the solvability of nonlinear equations with odd operators,* Funktsional. Anal. i Prilozhen. **1** (1967), no. 3, 66–73; English transl. in Functional Anal. Appl. **1** (1967).

[19] S. I. Pohozaev, *On an approach to nonlinear equations,* Dokl. Akad. Nauk SSSR **247** (1979), 1327–1331; English transl. in Soviet Math. Dokl. **20** (1979).

[20] S. I. Pohozaev, *On periodical solutions to some system at nonlinear ordinary diferential equations,* Differentsial'nye Uravneniya, 16, N 1, (1980), 109–116.

[21] S. I. Pohozaev, *On a constructive method in the calculus of variations,* Dokl. Akad. Nauk SSSR **298**, 1330–1333; English transl. in Soviet Math. Dokl. **37** (1988).

[22] S. I. Pohozaev, *On fibering method for the solution of nonlinear boundary value problems,* (in Russian) Trudy Math. Inst. of the Russian Academy of Sci. 192 (1990), 140–163. English. trans. Proceedings of the Steklov Institute of Mathematics, 1992, Issue 3, 157–173.

[23] S. I. Pohozaev, A. Tesei, *Existence of positive solutions of nonlinear Neumann problem* (to appear).

[24] P. Tolksdorf, *Regularity for a more general class of quasilinear elliptic equations,* J. Diff. Equations 51 (1984), 126–150.

[25] N. S. Trudinger, *On Harnack type inequalities and their application to quasilinear elliptic equations,* Comm. Pure Appl. Math. 20 (1967), 721–747.

Pohozaev Stanislav I.
Steklov Mathematical Institute
Vavilova 42
11766 Moscow, Russia

A SZULKIN

Index theories for indefinite functionals and applications

Preface

These notes are a somewhat expanded version of a series of lectures given by the author at the 20th Seminar in Partial Differential Equations held in Cholin in June 1995. I would like to thank the organizers for their kind invitation, for their hospitality and the perfect organization.

The material presented here is concerned with different index theories and their application to the problem of determining the number of periodic solutions to certain systems of ordinary differential equations. An index provides in a sense a measure for the size of subsets of a Banach space E which are invariant with respect to an action of some symmetry group. Using such index and a suitable minimax procedure it is possible to obtain a lower bound for the number of critical points of a functional Φ defined on E. Moreover, Φ and E may be chosen in such a way that critical points correspond to periodic solutions of our system of differential equations.

Although we assume that the reader has some basic knowledge of critical point theory and its applications to nonlinear differential equations, we usually recall necessary facts and give references for further reading. The results concerning the existence of periodic solutions to be found in these notes are not new. However, our approach - via limit index theories - is different from the usual one. It follows, with some modifications and extensions, the ideas developed by Y.Q. Li [30] and Fournier et al. [24].

1 Introduction

Let $V = V(x,t) : \mathbf{R}^N \times \mathbf{R} \to \mathbf{R}$ be a continuous function which is continuously differentiable in the x-variables and 2π-periodic in t. Consider the second order system of ordinary differential equations

$$-\ddot{x} = V_x(x,t) \tag{1.1}$$

with the periodic boundary conditions

$$x(0) - x(2\pi) = \dot{x}(0) - \dot{x}(2\pi) = 0. \tag{1.2}$$

The function V is called the *potential* because the solutions of (1.1) describe the Newtonian motion under the influence of the force field generated by V. An equivalent way of writing (1.1), (1.2) is

$$(1.3) \qquad -\ddot{x} = V_x(x,t), \qquad x(t+2\pi) = x(t).$$

To (1.3) (or (1.1), (1.2)) there corresponds the functional

$$(1.4) \qquad \Phi(x) := \frac{1}{2}\int_0^{2\pi} |\dot{x}|^2\,dt - \int_0^{2\pi} V(x,t)\,dt,$$

and it can be shown that critical points of Φ (in an appropriate space of 2π-periodic functions, to be defined later) are solutions of (1.3). Clearly, if $V(-x,t) = V(x,t)$ for all x, t, then the functional Φ is even.

A different situation occurs when V is independent of t, i.e. $V = V(x)$. The corresponding system is then called *autonomous* and one can look for solutions of any period T. Let $T = 2\pi$ and let T_θ be the phase-shift operator defined by $(T_\theta x)(t) := x(t+\theta)$. The functional Φ is then *invariant* in the sense that $\Phi(T_\theta x) = \Phi(x)$ for all x, θ. Since obviously $T_{\theta+2\pi} = T_\theta$, we may consider $\{T_\theta\}$ as a family of operators parametrized by S^1.

In the following sections we will assume that V_x is asymptotically linear at $x = 0$ and $|x| = \infty$ and either $V(-x,t) = V(x,t)$ for all x, t, or $V = V(x)$. The number of solutions of (1.3) will then be estimated by the difference between the Morse indices of the quadratic forms corresponding to the linearizations of (1.3) at 0 and ∞.

Consider now the Hamiltonian system of differential equations

$$(1.5) \qquad \begin{cases} \dot{p} = -H_q(p,q,t) \\ \dot{q} = H_p(p,q,t), \end{cases}$$

where $p, q \in \mathbf{R}^N$, and the *Hamiltonian* $H = H(p,q,t) : \mathbf{R}^{2N} \times \mathbf{R} \to \mathbf{R}$ is continuous, continuously differentiable in p, q and 2π-periodic in t. Note that if $H(p,q,t) = \frac{1}{2}|p|^2 + V(q,t)$ (i.e. H is the sum of the kinetic and the potential energy), then (1.5) reduces to $\dot{p} = -V_q$, $\dot{q} = p$ and again we obtain the second order system $-\ddot{q} = V_q(q,t)$. In the general case, set $z = (p,q) \in \mathbf{R}^{2N}$ and

$$J := \begin{pmatrix} 0 & -I \\ I & 0 \end{pmatrix},$$

where I is the $N \times N$ identity matrix. J is called the *standard symplectic* (or *skew-symmetric*) matrix and it has the obvious property $J^T = -J = J^{-1}$. With this notation (1.5) can be written in the more concise form $\dot{z} = JH_z(z,t)$. The periodic boundary value problem we want to study now is

$$\dot{z} = JH_z(z,t), \qquad z(0) = z(2\pi),$$

or equivalently,

$$(1.6) \qquad \dot{z} = JH_z(z,t), \qquad z(t+2\pi) = z(t).$$

To (1.6) there corresponds the functional

$$\Phi(z) := \frac{1}{2}\int_0^{2\pi} (-J\dot{z}\cdot z)\,dt - \int_0^{2\pi} H(z,t)\,dt, \tag{1.7}$$

and critical points of Φ solve (1.6). As before, we have that Φ is even if $H(-z,t) = H(z,t)$ for all z, t, and Φ is invariant with respect to the S^1-family of phase-shifts if $H = H(z)$ (the autonomous case). As we shall see later, the functional Φ is now *strongly indefinite* in the sense that it is unbounded below and above on any subspace of finite codimension and whenever the second derivative $\Phi''(z)$ exists, the quadratic form corresponding to it is positive definite and negative definite on infinite dimensional subspaces. The functional Φ corresponding to (1.3) is in general indefinite (i.e. unbounded from above and below) but is not *strongly* indefinite in the above sense.

Suppose that H_z is asymptotically linear at $z = 0$ and $|z| = \infty$. The Morse indices corresponding to the linearizations at 0 and ∞ are now infinite. However, it is still possible to introduce a number which in a sense is a relative Morse index and in particular is finite. The difference between these relative indices at 0 and ∞ will turn out to account for the number of solutions of (1.6).

We end this section by recalling some useful definitions and introducing some notation.

Let E be a real Banach space and let $\Phi \in C^1(E, \mathbf{R})$. The Fréchet derivative of Φ at x (which is an element of the dual space E^*) will be denoted by $\Phi'(x)$. A point $x_0 \in E$ is said to be *critical* if $\Phi'(x_0) = 0$. The corresponding value $c = \Phi(x_0)$ is called a *critical value* or a *critical level.* We will use the customary notation

$$K := \{x \in E : \Phi'(x) = 0\}, \qquad K_c := K \cap \Phi^{-1}(c)$$

and

$$\Phi^c := \{x \in E : \Phi(x) \le c\}.$$

Furthermore, we set

$$B_\rho := \{x \in E : \|x\| < \rho\} \qquad \text{and} \qquad \partial B_\rho := \{x \in E : \|x\| = \rho\}.$$

If $x \in E$ and $A \subset E$, then $d(x, A)$ denotes the distance from x to A. A sequence (x_j) is said to be a *Palais-Smale sequence* (or a *PS-sequence* for short) if $(\Phi(x_j))$ is bounded and $\Phi'(x_j) \to 0$. Φ satisfies the *Palais-Smale condition* (PS) if each PS-sequence has a convergent subsequence.

If E is a real Hilbert space with an inner product $\langle .\,, . \rangle$, then it will be convenient to define the *gradient* of Φ at x, $\nabla\Phi(x)$, by setting

$$\langle \nabla\Phi(x), y\rangle = \Phi'(x)y, \qquad y \in E.$$

Thus, while the Fréchet derivative is in the space E^*, the gradient is an element of E and $\Phi'(x)$ and $\nabla\Phi(x)$ are dual to each other.

Let G be a compact topological group. A continuous mapping $(g, x) \mapsto gx$ from $G \times E$ to E is called a *group action* if $ex = x$ for all $x \in E$ (where e is the unit element of G) and $(gh)x = g(hx)$ for all $g, h \in G$, $x \in E$. An action of G is *isometric* if for each $g \in G$ the mapping $x \mapsto gx$ is an isometry of E. A set $A \subset E$ is said to be *G-invariant* (or simply *invariant*) if $gA = A$ for all $g \in G$. A functional $\Phi : E \to \mathbf{R}$ is *G-invariant* if $\Phi(gx) = \Phi(x)$ for all g, x. A mapping $f : E_1 \to E_2$ between two Banach spaces is *G-equivariant* if $f(gx) = gf(x)$ for all g, x. Here the action of G on E_1 and E_2 need not be the same even if $E_1 = E_2$. The *fixed point set* of the action of G is defined as

$$E^G := \{x \in E : gx = x \text{ for all } g \in G\}.$$

The set $\mathcal{O}(x) := \{gx : g \in G\}$ is called the *orbit* of x. If the mappings $x \mapsto gx : E \to E$ are linear for each $g \in G$, it may be convenient to denote them by T_g and write $T_g x$ instead of gx. Observe that in such case E^G is necessarily a subspace of E. The family $\mathbf{T} = \{T_g\}_{g \in G}$ will also be called a *representation* of G in E (an *isometric representation* if all T_g are isometries).

Index theories for general group actions and their applications may be found in the monograph [6]. Here we will only be concerned with the two simplest cases: $G = \mathbf{Z}_2$ and $G = S^1$.

For $G = \mathbf{Z}_2 = \{0, 1\}$ we only consider the (isometric) group action on E given by $T_0 = id$, $T_1 = -id$ (id is the identity mapping). Then a set $A \subset E$ is invariant if and only if $-A = A$, invariant functionals are the ones that are even, and equivariant mappings are those which are odd. Moreover, $E^G = \{0\}$. Note that if V is even in x and H even in z, then the functionals Φ given by (1.4) and (1.7) are invariant with respect to this action of $\mathbf{Z}_2$.

Let now $G = S^1 \equiv \mathbf{R}/2\pi\mathbf{Z}$. If $z = (z_1, \ldots, z_k) \in \mathbf{C}^k$ and $n_1, \ldots, n_k \in \mathbf{Z}$, then $\mathbf{S} = \{S_\theta\}_{\theta \in S^1}$, where $S_\theta z := (e^{in_1\theta} z_1, \ldots, e^{in_k\theta} z_k)$, is an isometric representation of S^1 in $\mathbf{C}^k$. Note that $E^G = \{0\}$ if and only if all $n_j \neq 0$. If E is some Banach space of 2π-periodic functions, then the family $\mathbf{T} = \{T_\theta\}_{\theta \in S^1}$, where T_θ is the phase-shift operator introduced earlier, is an isometric representation of S^1 in E. As we have already seen, if $V = V(x)$ and $H = H(z)$, then the functionals Φ given by (1.4) and (1.7) are invariant with respect to $\mathbf{T}$. Note that here E^G is the subspace of constant functions. The fact that $E^G \neq \{0\}$ will turn out to make the search of critical points much more difficult.

2 Genus and minimax principles for even functionals

Let E be a real Banach space and

$$\Sigma := \{A \subset E : A \text{ is closed and } -A = A\}.$$

Definition 2.1 Let $A \in \Sigma$, $A \neq \emptyset$. The *genus* of A, denoted $\gamma(A)$, is the smallest integer k such that there exists an odd mapping $f \in C(A, \mathbf{R}^k - \{0\})$. If such mapping does not exist for any k, then $\gamma(A) := +\infty$. Finally, $\gamma(\emptyset) = 0$.

The notion of genus is due to Krasnoselskii [25]. The above definition is that of Coffman [16] and it differs from Krasnoselskii's original one. The proof that the two definitions are equivalent has been established by Rabinowitz [36].

Observe that if $0 \in A$, then $\gamma(A) = +\infty$ and if $\gamma(A) \geq 2$, then either $0 \in A$ or A is an infinite set. Let us also remark that some authors do not define the genus for sets $A \in \Sigma$ which contain the origin.

Properties of genus are summarized in the following

Proposition 2.2 *Let* $A, B \in \Sigma$.
(i) If there exists an odd mapping $g \in C(A, B)$, *then* $\gamma(A) \leq \gamma(B)$.
(ii) $\gamma(A \cup B) \leq \gamma(A) + \gamma(B)$.
(iii) If A *is compact, then there exists a neighbourhood* $N \in \Sigma$ *of* A *such that* $\gamma(A) = \gamma(N)$. *Moreover,* $\gamma(A) < \infty$ *if* $0 \notin A$.
(iv) $\gamma(S^{l-1}) = l$, *where* S^{l-1} *is the unit sphere in* $\mathbf{R}^l$.
(v) If X *is a subspace of codimension* m *in* E *and* $\gamma(A) > m$, *then* $A \cap X \neq \emptyset$.

These properties are well-known and their proofs may be found e.g. in [38, Section 7] or [39, Section II.5]. For the sake of illustration, let us prove (v). Suppose that $A \cap X = \emptyset$, write $E = X \oplus Y$, and for $x + y \in X \oplus Y$, let $f(x+y) = y$. Then f is an odd mapping and $f|_A \in C(A, Y - \{0\})$. Since $\dim Y = m$, $\gamma(A) \leq m$, a contradiction.

Theorem 2.3 *Suppose* $\Phi \in C^1(E, \mathbf{R})$ *is even, satisfies (PS),* $\Phi(0) = 0$ *and there exist numbers* $\rho > 0$, $\alpha < \beta < 0$ *and subspaces* X, Y *of* E *such that:*
(i) $m = \operatorname{codim} X < \dim Y = l < \infty$;
(ii) $\Phi|_{Y \cap \partial B_\rho} \leq \beta$;
(iii) $\Phi|_X \geq \alpha$.
Then Φ *has at least* $l - m$ *pairs of critical points* $\pm x_j$. *The corresponding critical values can be characterized as*

$$c_j = \inf_{\gamma(A) \geq j} \sup_{x \in A} \Phi(x), \qquad m + 1 \leq j \leq l,$$

and are contained in the interval $[\alpha, \beta]$.

Theorem 2.3 is due to Clark [14], see also [38, Theorem 9.1]. An important role in the proof of this theorem will be played by the *deformation lemma.* A mapping $\eta : A \times [0,1] \to E$, where $A \subset E$, is called a *deformation* (or more precisely a *deformation of* A *in* E) if $\eta(x, 0) = x$ for all $x \in A$.

Lemma 2.4 *Suppose $\Phi \in C^1(E, \mathbf{R})$ and let $c \in \mathbf{R}$, $\bar{\varepsilon}, \delta > 0$ and a set $N \subset E$ be given. If*

$$(2.1) \qquad \|\Phi'(x)\| \geq \delta \quad \text{whenever } d(x, E - N) \leq \delta \text{ and } |\Phi(x) - c| \leq \bar{\varepsilon},$$

then there exists an $\varepsilon \in (0, \bar{\varepsilon})$, depending only on $\bar{\varepsilon}$ and δ, and a deformation $\eta : E \times [0,1] \to E$ such that:
(i) $\eta(x,t) = x$ whenever $|\Phi(x) - c| \geq \bar{\varepsilon}$;
(ii) $\eta(\Phi^{c+\varepsilon} - N, 1) \subset \Phi^{c-\varepsilon}$ and $\eta(\Phi^{c+\varepsilon}, 1) \subset \Phi^{c-\varepsilon} \cup N$;
(iii) the mapping $t \mapsto \Phi(\eta(x,t))$ is nonincreasing for each $x \in E$;
(iv) the mapping $x \mapsto \eta(x,t)$ is odd for each t if Φ is even and $-N = N$.

The above result is a version of the quantitative deformation lemma due to Willem [44], [45, Lemma 2.3], see also Chang [13, Theorem I.3.4]. We emphasise here that the constant ε does not depend on the functional Φ (or the space E) as long as Φ satisfies (2.1). More common (non-quantitative) versions of this lemma may be found e.g. in [34, 38, 39]. Let us remark that such versions would be sufficient for the proof of Theorem 2.3. However, later on we will need the extra information contained in Lemma 2.4 that ε is independent of the particular choice of the functional Φ.

Proof of Lemma 2.4 Since the proof is very similar to that of the usual deformation lemma, we only sketch the argument.

A mapping $V : E - K \to E$ is said to be a *pseudo-gradient vector field* for Φ if V is locally Lipschitz continuous and satisfies

$$(2.2) \qquad \|V(x)\| \leq 2\|\Phi'(x)\|, \qquad \Phi'(x)V(x) \geq \|\Phi'(x)\|^2$$

for each $x \in E - K$. It is well-known that if $\Phi \in C^1(E, \mathbf{R})$, then there exists a pseudo-gradient vector field for Φ [34, Lemma 6.1], [38, Lemma A.2], [39, Lemma II.3.2]. Moreover, if Φ is even, $V(x)$ may be replaced by $\frac{1}{2}(V(x) - V(-x))$, so we may assume that V is odd.

Let $\chi : E \to [0,1]$ be a locally Lipschitz continuous function such that

$$\chi(x) = \begin{cases} 0 & \text{if } |\Phi(x) - c| \geq \bar{\varepsilon} \text{ or } d(x, E - N) \geq \delta \\ 1 & \text{if } |\Phi(x) - c| \leq \bar{\varepsilon}/2 \text{ and } d(x, E - N) \leq \delta/2 \end{cases}$$

and consider the Cauchy problem

$$\frac{d\eta}{dt} = -\frac{1}{2}\delta\chi(\eta(x,t))\frac{V(\eta(x,t))}{\|V(\eta(x,t))\|}, \qquad \eta(x,0) = x.$$

Since the vector field above is locally Lipschitz continuous and bounded, $\eta(x,t)$ is uniquely determined and continuous for each $(x,t) \in E \times \mathbf{R}$. It is now easily seen that (i) and (iv) are satisfied (note that we can choose an even χ if Φ is even and $-N = N$). Moreover,

$$(2.3) \qquad \frac{d}{dt}\Phi(\eta(x,t)) = \Phi'(\eta(x,t))\frac{d\eta}{dt} \leq -\frac{\delta}{4}\chi(\eta(x,t))\|\Phi'(\eta(x,t))\|$$

according to (2.2). Hence also (iii) holds.

Let $x \in \Phi^{c+\varepsilon} - N$ and $0 < \varepsilon \le \overline{\varepsilon}/2$. In order to establish the first part of (ii) we must show that $\Phi(\eta(x,1)) \le c - \varepsilon$. Since

$$\|\eta(x,t) - x\| \le \int_0^t \left\| \frac{d\eta}{ds} \right\| ds \le \frac{1}{2}\delta t,$$

$d(\eta(x,t), E - N) \le \delta/2$ whenever $0 \le t \le 1$. We may assume $\Phi(\eta(x,1)) \ge c - \overline{\varepsilon}/2$ (otherwise we are done). Then, according to (2.3) and the definition of χ,

$$\Phi(\eta(x,1)) = \Phi(x) + \int_0^1 \frac{d}{dt}\Phi(\eta(x,t))\, dt \le \Phi(x) - \frac{\delta}{4}\int_0^1 \|\Phi'(\eta(x,t))\|\, dt \le c + \varepsilon - \delta^2/4.$$

Hence $\Phi(\eta(x,1)) \le c - \varepsilon$ if we choose $\varepsilon \le \min\{\overline{\varepsilon}/2, \delta^2/8\}$.

In order to prove the second part of (ii) it remains to observe that if $x \in \Phi^{c+\varepsilon}$ and $\eta(x,t) \notin N$, then $d(\eta(x,t), E - N) \le \delta/2$ and therefore again $\Phi(\eta(x,1)) \le c - \varepsilon$. □

Proof of Theorem 2.3 Since $\{A \in \Sigma : \gamma(A) \ge j+1\} \subset \{A \in \Sigma : \gamma(A) \ge j\}$, $c_{m+1} \le c_{m+2} \le \ldots \le c_l$. Since $\gamma(Y \cap \partial B_\rho) = l$ according to (iv) of Proposition 2.2, $c_l \le \sup_{Y \cap \partial B_\rho} \Phi \le \beta$. If $\gamma(A) \ge m+1$, then $A \cap X \ne \emptyset$ by (v) of Proposition 2.2, so $\sup_A \Phi \ge \alpha$ and $c_{m+1} \ge \alpha$. Thus all c_j are contained in the interval $[\alpha, \beta]$.

Suppose $c := c_j = \ldots = c_{j+p}$ for some $p \ge 0$. Since $c < 0$, $0 \notin K_c$. So the proof will be complete if we can show that $\gamma(K_c) \ge p+1$ (because either all c_j are distinct and $K_{c_j} \ne \emptyset$ or $\gamma(K_{c_j}) \ge 2$ for some j and K_{c_j} is an infinite set). It follows from (PS) that K_c is compact. There exists a neighbourhood $N \in \Sigma$ of K_c such that $\gamma(K_c) = \gamma(N)$. We can find $\overline{\varepsilon}, \delta > 0$ such that (2.1) is satisfied. For otherwise there is a sequence (x_j) with $d(x_j, E - N) \to 0$, $\Phi(x_j) \to c$ and $\Phi'(x_j) \to 0$. Passing to a subsequence and using (PS) we may assume that $x_j \to \overline{x} \in K_c$. However, since $d(x_j, E - N) \to 0$ and K_c is compact, $\overline{x}$ is bounded away from K_c. This contradiction shows that $\overline{\varepsilon}$ and δ as above exist. Let now ε and η be as in the conclusion of Lemma 2.4. Then $\eta(.,1)$ is odd and maps $\Phi^{c+\varepsilon}$ into $\Phi^{c-\varepsilon} \cup N$. Since

$$c = \inf_{\gamma(A) \ge j+p} \sup_{x \in A} \Phi(x) = \inf_{\gamma(A) \ge j} \sup_{x \in A} \Phi(x),$$

$\gamma(\Phi^{c+\varepsilon}) \ge j + p$ and $\gamma(\Phi^{c-\varepsilon}) < j$. It follows therefore from (i) and (ii) of Proposition 2.2 that

$$j + p \le \gamma(\Phi^{c+\varepsilon}) \le \gamma(\Phi^{c-\varepsilon} \cup N) \le \gamma(\Phi^{c-\varepsilon}) + \gamma(N) < j + \gamma(N)$$

and $\gamma(K_c) = \gamma(N) > p$. □

Let $(E_n)_{n=1}^{\infty}$ be a sequence of closed subspaces of E such that $E_n \subset E_{n+1}$ for each $n \ge 1$ and

(2.4) $$E = \overline{\bigcup_{n=1}^{\infty} E_n}.$$

Let $(d_n)_{n=1}^{\infty}$ be a sequence of integers and

$$\mathcal{E} := \{E_n, d_n\}_{n=1}^{\infty}.$$

In applications we will either have $\dim E_n = b + kn$ and $d_n = c + dn$ for some $b, c \geq 0$, $0 < d \leq k$, or $E_n = E$ and $d_n = 0$ for all n.

Definition 2.5 Let $A \in \Sigma$. The *limit genus* of A *with respect* to $\mathcal{E}$, $\gamma_{\mathcal{E}}(A)$, is defined by

$$\gamma_{\mathcal{E}}(A) := \limsup_{n\to\infty}(\gamma(A \cap E_n) - d_n).$$

The notion of limit genus (in a somewhat different form) is due to Y.Q. Li [30]. A related concept of limit category may be found in Fournier et al. [24].

Note that for each $A \in \Sigma$, $\gamma_{\mathcal{E}}(A) \in \mathbf{Z} \cup \{-\infty, +\infty\}$. If $E_n = E$ and $d_n = 0$ for all n, then $\gamma_{\mathcal{E}} \equiv \gamma$. Since we will often encounter sets of the form $A \cap E_n$, it will be convenient to adopt the notation

$$A_n := A \cap E_n.$$

Proposition 2.6 *Let* $A, B \in \Sigma$.
(i) If for almost all n *there exists an odd mapping* $g_n \in C(A_n, B_n)$, *then* $\gamma_{\mathcal{E}}(A) \leq \gamma_{\mathcal{E}}(B)$.
(ii) $\gamma_{\mathcal{E}}(A \cup B) \leq \gamma_{\mathcal{E}}(A) + \gamma(B)$.
(iii) Let $l \in \mathbf{Z}$, $R > 0$. *If* Y *is a subspace of* E *such that* $\dim Y_n = d_n + l$ *for almost all* n, *then* $\gamma_{\mathcal{E}}(Y \cap \partial B_R) = l$.
(iv) Let $m \in \mathbf{Z}$. *If* X *is a subspace of* E *such that* $\operatorname{codim}_{E_n} X_n = d_n + m$ *for almost all* n *and if* $\gamma_{\mathcal{E}}(A) > m$, *then* $A \cap X \neq \emptyset$.

Proof We will use Proposition 2.2 and Definition 2.5.

(i) For almost all n, $\gamma(A_n) - d_n \leq \gamma(B_n) - d_n$. Now it suffices to pass to the limit as $n \to \infty$.

(ii) $\gamma(A_n \cup B_n) - d_n \leq \gamma(A_n) - d_n + \gamma(B_n) \leq (\gamma(A_n) - d_n) + \gamma(B)$.

(iii) For almost all n, $\gamma(Y_n \cap \partial B_R) = d_n + l$.

(iv) There exists a number n such that $\operatorname{codim}_{E_n} X_n = d_n + m$ and $\gamma(A_n) > d_n + m$. Therefore $\emptyset \neq A_n \cap X_n \subset A \cap X$. □

Let

$$\Phi_n := \Phi|_{E_n}.$$

For each $x \in E_n$, $\Phi_n'(x) \in E_n^*$.

Definition 2.7 A sequence (x_j) is said to be a *PS*-sequence* for Φ (with respect to $\mathcal{E}$) if $(\Phi(x_j))$ is bounded, each $x_j \in E_{n_j}$ for some n_j, $n_j \to \infty$ and $\Phi_{n_j}'(x_j) \to 0$ as $j \to \infty$. Φ is said to satisfy the *(PS)*-condition* (with respect to $\mathcal{E}$) if each PS*-sequence has a subsequence converging to a critical point.

Condition (PS)* (in a slightly different form) has been introduced independently by Bahri and Berestycki [3, 4] and Li and Liu [27]. It is easy to see that if Φ satisfies (PS)*, then K_c is compact for each c. Moreover, Φ satisfies (PS). Indeed, suppose (x_j) is a sequence such that $\Phi(x_j)$ is bounded and $\Phi'(x_j) \to 0$. It follows from (2.4) that for each j there are $n_j \geq j$ and $y_j \in E_{n_j}$ such that $|\Phi(x_j) - \Phi(y_j)| \leq 1/j$, $\|y_j - x_j\| \leq 1/j$ and $\|\Phi'(x_j) - \Phi'(y_j)\| \leq 1/j$. Since $\|\Phi'_{n_j}(y_j)\| \leq \|\Phi'(y_j)\|$, $\Phi'_{n_j}(y_j) \to 0$. So (y_j), and therefore also (x_j), has a convergent subsequence.

Theorem 2.8 *Suppose $\Phi \in C^1(E, \mathbf{R})$ is even, satisfies (PS)*, $\Phi(0) = 0$ and there exist numbers $\rho > 0$, $\alpha < \beta < 0$ and subspaces X, Y of E such that:*
(i) $\operatorname{codim}_{E_n} X_n = d_n + m$ and $\dim Y_n = d_n + l$ for almost all n and some $m, l \in \mathbf{Z}$, $m < l$;
(ii) $\Phi|_{Y \cap \partial B_\rho} \leq \beta$;
(iii) $\Phi|_X \geq \alpha$.
Then Φ has at least $l - m$ pairs of critical points $\pm x_j$. The corresponding critical values can be characterized as

$$c_j = \inf_{\gamma_{\mathcal{E}}(A) \geq j} \sup_{x \in A} \Phi(x), \qquad m + 1 \leq j \leq l,$$

and are contained in the interval $[\alpha, \beta]$.

Proof As in the proof of Theorem 2.3, we see that $c_{m+1} \leq c_{m+2} \leq \ldots \leq c_l$. Since $\dim Y_n = d_n + l$ for almost all n, $\gamma_{\mathcal{E}}(Y \cap \partial B_\rho) = l$ and $c_l \leq \beta$. If $\gamma_{\mathcal{E}}(A) \geq m + 1$, then $A \cap X \neq \emptyset$. So $c_{m+1} \geq \alpha$.

Let $c := c_j = \ldots = c_{j+p}$ for some $p \geq 0$. It remains to show that $\gamma(K_c) \geq p + 1$. Choose a neighbourhood $N \in \Sigma$ of K_c with $\gamma(N) = \gamma(K_c)$. We claim that there are $\bar{\varepsilon}, \delta > 0$ and $n_0 \geq 1$ such that

(2.5) $\quad \|\Phi'_n(x)\| \geq \delta \quad$ whenever $d(x, E_n - N) \leq \delta$, $|\Phi_n(x) - c| \leq \bar{\varepsilon}$ and $n \geq n_0$.

Assuming the contrary we can find a sequence (x_j) such that $x_j \in E_{n_j}$ for some $n_j \geq j$, $d(x_j, E_{n_j} - N) \to 0$, $\Phi_{n_j}(x_j) \to c$ and $\Phi'_{n_j}(x_j) \to 0$. So (x_j) is a PS*-sequence. Passing to a subsequence, we obtain that $x_j \to \bar{x} \in K_c$ and $\bar{x}$ is bounded away from K_c, a contradiction (cf. the proof of Theorem 2.3).

Now it follows from Lemma 2.4 that we can find an $\varepsilon > 0$ such that for each $n \geq n_0$ there exists a deformation $\eta_n : E_n \times [0, 1] \to E_n$ with $\eta_n(\Phi_n^{c+\varepsilon}, 1) \subset \Phi_n^{c-\varepsilon} \cup N_n$. So by Proposition 2.6 and the definition of c,

$$j + p \leq \gamma_{\mathcal{E}}(\Phi^{c+\varepsilon}) \leq \gamma_{\mathcal{E}}(\Phi^{c-\varepsilon} \cup N) \leq \gamma_{\mathcal{E}}(\Phi^{c-\varepsilon}) + \gamma(N) < j + \gamma(N).$$

Hence $\gamma(K_c) = \gamma(N) > p$. □

Theorem 2.8 is a variant of a result contained in [30]. Earlier results of similar type may be found e.g. in Benci [9]. Let us remark that while in [9] E is a Hilbert space, Φ satisfies (PS) and Φ' is of the form *linear + compact*, here we assume that Φ satisfies condition (PS)* - which is stronger than (PS) - but on the other hand we need no extra assumption on Φ' (and E need not be a Hilbert space).

Although Theorem 2.8 is designed to treat strongly indefinite functionals like those considered in Section 4, it may sometimes be convenient to use it for functionals which are not strongly indefinite, see Section 3. However, in this latter case a stronger result (by a more technical proof) has been obtained by Bartolo et al. [5, Theorem 2.4].

Remark 2.9 If E is a Hilbert space and (x_j) a PS*-sequence converging to $\overline{x}$, then $\overline{x}$ is necessarily a critical point. Indeed, let $P_n : E \to E_n$ be the orthogonal projector. Then

$$\langle \nabla\Phi(x_j) - \nabla\Phi_{n_j}(x_j), z\rangle = \langle \nabla\Phi(x_j) - P_{n_j}\nabla\Phi(x_j), z\rangle = \langle \nabla\Phi(x_j), z - P_{n_j}z\rangle \to 0$$

for each $z \in E$. Since $\nabla\Phi(x_j) \to \nabla\Phi(\overline{x})$ and $\nabla\Phi_{n_j}(x_j) \to 0$, we see that $\nabla\Phi(\overline{x}) = 0$.

Remark 2.10 Theorem 2.3 is a special case of Theorem 2.8 (take $\mathcal{E} = \{E_n, d_n\}$ with $E_n = E$ and $d_n = 0$ for all n).

3 Second order systems with even potential

In this section we will be concerned with the problem

$$-\ddot{x} = V_x(x,t), \qquad x(t+2\pi) = x(t), \tag{3.1}$$

where $V \in C(\mathbf{R}^N \times \mathbf{R}, \mathbf{R})$, $V_x \in C(\mathbf{R}^N \times \mathbf{R}, \mathbf{R}^N)$ and V is 2π-periodic in t. Let $E := H^1(S^1, \mathbf{R}^N)$ be the Sobolev space of 2π-periodic $\mathbf{R}^N$-valued functions

$$x(t) = a_0 + \sum_{k=1}^{\infty} a_k \cos kt + b_k \sin kt, \qquad a_0, a_k, b_k \in \mathbf{R}^N \tag{3.2}$$

such that $\sum_{k=1}^{\infty} k^2(|a_k|^2 + |b_k|^2) < \infty$. Then E is a Hilbert space with an inner product given by

$$\langle x, x'\rangle := 2\pi a_0 \cdot a_0' + \pi \sum_{k=1}^{\infty} k^2 (a_k \cdot a_k' + b_k \cdot b_k'). \tag{3.3}$$

By the Parseval equality we have

$$\int_0^{2\pi} (|x(t)|^2 + |\dot{x}(t)|^2)\, dt = 2\pi |a_0|^2 + \pi \sum_{k=1}^{\infty} (1+k^2)(|a_k|^2 + |b_k|^2),$$

so the norm in E corresponding to the inner product (3.3) is equivalent to the usual Sobolev norm. Recall that the space E is compactly embedded in $C(S^1, \mathbf{R}^N)$. Set

$$\Phi(x) := \frac{1}{2}\int_0^{2\pi} |\dot{x}|^2\, dt - \int_0^{2\pi} V(x,t)\, dt.$$

Then $\Phi \in C^1(E, \mathbf{R})$ and $\nabla\Phi(x) = 0$ if and only if $x = x(t)$ is a solution of (3.1) [34, Section 1.4], [38, Appendix B].

Let now A be a symmetric $N \times N$ constant matrix and consider the problem

$$-\ddot{x} = Ax, \qquad x(t+2\pi) = x(t). \tag{3.4}$$

Let $L : E \to E$ be the linear operator given by

$$\langle Lx, y\rangle := \int_0^{2\pi} (\dot{x} \cdot \dot{y} - Ax \cdot y)\, dt. \tag{3.5}$$

Clearly, $Lx = 0$ if and only if x satisfies (3.4). Substituting (3.2) into (3.5) with $y = x$ gives

$$\langle Lx, x\rangle = -2\pi A a_0 \cdot a_0 + \pi \sum_{k=1}^{\infty} \left[(k^2 I - A)a_k \cdot a_k + (k^2 I - A)b_k \cdot b_k\right]. \tag{3.6}$$

After a change of coordinates we may assume that $A = \operatorname{diag}(\mu_1, \ldots, \mu_N)$. Recall that the Morse index of a quadratic form is the maximal dimension of a subspace on which this form is negative definite. So e.g. the form $a_0 \mapsto -2\pi A a_0 \cdot a_0$ in $\mathbf{R}^N$ has the index equal to $\#\{j : \mu_j > 0\}$.

Denote the Morse index of the quadratic form (3.6) by $i^-(A)$ and the nullity (i.e. $\dim N(L)$) by $i^0(A)$. A simple computation shows that

$$\begin{aligned} i^-(A) &= \#\{j : \mu_j > 0\} + 2\sum_{k=1}^{\infty} \#\{j : \mu_j > k^2\} \\ &= \#\{j : \mu_j > 0\} + 2\sum_{j=1}^{N} \#\{k \geq 1 : \mu_j > k^2\} \end{aligned} \tag{3.7}$$

and

$$\begin{aligned} i^0(A) &= \#\{j : \mu_j = 0\} + 2\sum_{k=1}^{\infty} \#\{j : \mu_j = k^2\} \\ &= \#\{j : \mu_j = 0\} + 2\sum_{j=1}^{N} \#\{k \geq 1 : \mu_j = k^2\}. \end{aligned} \tag{3.8}$$

One sees that $i^-(A)$ is finite. Since $i^0(A)$ is the number of linearly independent solutions of (3.4), $i^0(A) \leq 2N$. Moreover, $i^0(A) = 0$ (and L is invertible) if and only if $\sigma(A) \cap \{k^2 : k = 0, 1, 2, \ldots\} = \emptyset$, where $\sigma(A)$ denotes the spectrum of A.

Suppose now that there exist two symmetric $N \times N$ constant matrices A_0 and A_∞ such that

$$V(x,t) = \frac{1}{2} A_0 x \cdot x + W_0(x,t), \tag{3.9}$$

where $(W_0)_x(x,t) = o(|x|)$ uniformly in t as $x \to 0$ and

$$V(x,t) = \frac{1}{2} A_\infty x \cdot x + W_\infty(x,t), \tag{3.10}$$

where $(W_\infty)_x(x,t) = o(|x|)$ uniformly in t as $|x| \to \infty$.

Clearly, if (3.9) is satisfied, then $x = 0$ is a solution of (3.1) (it will be referred to as the *trivial* solution).

Theorem 3.1 *Suppose $V \in C(\mathbf{R}^N \times \mathbf{R}, \mathbf{R})$, $V_x \in C(\mathbf{R}^N \times \mathbf{R}, \mathbf{R}^N)$, V is 2π-periodic in t and $V(-x,t) = V(x,t)$ for all x, t. Suppose also V satisfies (3.9), (3.10) and $\sigma(A_\infty) \cap \{k^2 : k = 0, 1, 2, \ldots\} = \emptyset$.*
(i) If $i^-(A_0) > i^-(A_\infty)$, then (3.1) has at least $i^-(A_0) - i^-(A_\infty)$ pairs of nontrivial solutions.
(ii) If $i^-(A_0) + i^0(A_0) < i^-(A_\infty)$, then (3.1) has at least $i^-(A_\infty) - i^-(A_0) - i^0(A_0)$ pairs of nontrivial solutions.

The above result (in a more general form) is due to Clark [15]. See also Remark 3.4 below. In order to prove Theorem 3.1 we will need some preliminaries. It will be convenient to use the following notation:

$$(3.11)\quad \begin{aligned}\Phi(x) &= \frac{1}{2}\int_0^{2\pi} (|\dot{x}|^2 - A_0 x \cdot x)\, dt - \int_0^{2\pi} W_0(x,t)\, dt =: \frac{1}{2}\langle L_0 x, x\rangle - \varphi_0(x) \\ &= \frac{1}{2}\int_0^{2\pi} (|\dot{x}|^2 - A_\infty x \cdot x)\, dt - \int_0^{2\pi} W_\infty(x,t)\, dt =: \frac{1}{2}\langle L_\infty x, x\rangle \varphi_\infty(x).\end{aligned}$$

Lemma 3.2 *If $V \in C(\mathbf{R}^N \times \mathbf{R}, \mathbf{R})$, $V_x \in C(\mathbf{R}^N \times \mathbf{R}, \mathbf{R}^N)$, V is 2π-periodic in t, $V(0,t) = 0$ for all t and (3.9), (3.10) are satisfied, then $\nabla\varphi_0(x) = o(\|x\|)$, $\varphi_0(x) = o(\|x\|^2)$ as $x \to 0$ and $\nabla\varphi_\infty(x) = o(\|x\|)$, $\varphi_\infty(x) = o(\|x\|^2)$ as $\|x\| \to \infty$.*

For the proof, see Lemma 4.2, where a somewhat more difficult situation is considered.

Let

$$E_n := \{x \in E : x(t) = a_0 + \sum_{k=1}^{n} a_k \cos kt + b_k \sin kt\},$$

$d_n := \dim E_n = N(1 + 2n)$ and $\mathcal{E} := \{E_n, d_n\}_{n=1}^{\infty}$. Note that L_0 and L_∞ map E_n into itself.

Lemma 3.3 *Suppose $V \in C(\mathbf{R}^N \times \mathbf{R}, \mathbf{R})$, $V_x \in C(\mathbf{R}^N \times \mathbf{R}, \mathbf{R}^N)$, V is 2π-periodic in t, satisfies (3.10) and $\sigma(A_\infty) \cap \{k^2 : k = 0, 1, 2, \ldots\} = \emptyset$. Then Φ satisfies (PS) and (PS)*.*

Proof Denote the orthogonal projector of E onto E_n by P_n. Let (x_j) be a PS*-sequence. Then

$$(3.12)\qquad \nabla\Phi_{n_j}(x_j) = P_{n_j}\nabla\Phi(x_j) = L_\infty x_j - P_{n_j}\nabla\varphi_\infty(x_j) \to 0.$$

Since L_∞ is invertible and $\nabla\varphi_\infty(x) = o(\|x\|)$ as $\|x\| \to \infty$, (x_j) is bounded. So we may assume after passing to a subsequence that $x_j \to \overline{x}$ weakly in E and strongly in

$C(S^1, \mathbf{R}^N)$. Therefore $P_{n_j}\nabla\varphi_\infty(x_j) \to \nabla\varphi_\infty(\overline{x})$ (cf. [34, Section 1.4], [38, Appendix B]) and it follows from (3.12) that $x_j \to \overline{x}$. Hence Φ satisfies (PS)* (that $\nabla\Phi(\overline{x}) = 0$ is seen either from Remark 2.9 or directly from (3.12)).

The proof that Φ satisfies (PS) is similar. □

Let us introduce the notation

$$E = E^+(L) \oplus E^0(L) \oplus E^-(L),$$

where $E^+(L)$, $E^0(L)$, $E^-(L)$ are the L-invariant subspaces corresponding to the positive, the zero and the negative part of the spectrum of L.

Proof of Theorem 3.1 We may assume without loss of generality that $V(0,t) = 0$ for all t. So $\Phi(0) = 0$. Since $V(-x,t) = V(x,t)$ for all x, t, the functional Φ is even.

(i) We will use Theorem 2.3. It follows from Lemma 3.3 that Φ satisfies (PS). Let $l := i^-(A_0)$, $m := i^-(A_\infty)$, $X := E^+(L_\infty)$ and $Y := E^-(L_0)$. Then $\dim Y = l$. Since $E = E^+(L_\infty) \oplus E^-(L_\infty)$ and $\dim E^-(L_\infty) = m$, $\operatorname{codim} X = m$. Furthermore, $\langle L_\infty x, x\rangle \geq \varepsilon\|x\|^2$ for some $\varepsilon > 0$ and all $x \in X$. Invoking Lemma 3.2 we see that Φ is bounded below on X. This gives (iii) of Theorem 2.3. Finally, $\langle L_0 x, x\rangle \leq -\varepsilon\|x\|^2$ for some $\varepsilon > 0$ and all $x \in Y$. So using Lemma 3.2 once more we see that if $\rho > 0$ is small enough, then $\Phi|_{Y\cap\partial B_\rho} \leq \beta$ for a suitably chosen $\beta < 0$. This completes the verification of the hypotheses of Theorem 2.3.

(ii) Here we will apply Theorem 2.8 to the functional $-\Phi$. Note that $-\Phi$ satisfies (PS)* (since Φ does). Let now $\widetilde{l} := i^-(A_0) + i^0(A_0)$, $\widetilde{m} := i^-(A_\infty)$, $X := E^+(-L_\infty) \equiv E^-(L_\infty)$ and $Y := E^-(-L_0) \equiv E^+(L_0)$. We see as in the proof of (i) that $-\Phi|_X \geq \alpha$ and $-\Phi|_{Y\cap\partial B_\rho} \leq \beta$ for suitably chosen α, β and ρ ($\alpha < \beta < 0$, $\rho > 0$). If n is large enough, then $E^-(L_0)$, $E^0(L_0)$ and $E^-(L_\infty)$ are contained in E_n. Hence for such n, $\dim Y_n = \dim E_n - \dim E^-(L_0) - \dim E^0(L_0) = d_n - \widetilde{l}$. Similarly, $\operatorname{codim}_{E_n} X_n = \dim E_n - \dim E^-(L_\infty) = d_n - \widetilde{m}$. So according to Theorem 2.8, (3.1) has at least $-\widetilde{l} - (-\widetilde{m}) = \widetilde{m} - \widetilde{l}$ pairs of nontrivial solutions. □

Remark 3.4 (i) If the matrices A_0 and A_∞ in (3.9), (3.10) are nonconstant and have 2π-periodic entries, then the Morse index i^- and the nullity i^0 are still finite numbers but they are no longer given by (3.7), (3.8). The conclusions of Theorem 3.1 remain valid in this case. The proof of (i) is unchanged while the proof of (ii) would require some additional work (in order to verify (PS)* and to compute the dimensions of X_n and Y_n). We do not carry out the details and refer to [15] where a different argument is given or to [5, Theorem 2.4] from which the conclusion of (ii) follows directly.

(ii) If L_∞ is not invertible, $(W_\infty)_x$ is bounded and $W_\infty(x,t) \to \infty$ (or $-\infty$) uniformly in t as $|x| \to \infty$, then one can show that Φ satisfies (PS) and (PS)* (cf. the proof of Theorem 4.1 in [8]) and a result similar to Theorem 3.1 can be obtained. If $W_0(x,t)$ has constant sign for some $\delta > 0$ and all $0 < |x| < \delta$, then in some cases the subspace Y in the proof of Theorem 3.1 can be replaced by a larger one and a somewhat better result is valid. E.g., if $(W_\infty)_x$ is bounded, $W_\infty(x,t) \to \infty$ uniformly in t as $|x| \to \infty$ and $W_0(x,t) < 0$ for $0 < |x| < \delta$, then we can choose $X = E^-(L_\infty) \oplus E^0(L_\infty)$ and $Y = E^+(L_0) \oplus E^0(L_0)$ in the proof of (ii). So (3.1) has at least $i^-(A_\infty) + i^0(A_\infty) - i^0(A_0)$ pairs of nontrivial solutions in this case. On the other hand, if $(W_\infty)_x$ is bounded, $W_\infty(x,t) \to -\infty$ uniformly in t as $|x| \to \infty$ and $W_0(x,t) > 0$ for $0 < |x| < \delta$, then we must take $X = E^-(L_\infty)$, $Y = E^+(L_0)$, and we obtain the same conclusion as in (ii) of Theorem 3.1 (in fact here the condition $W_0(x,t) > 0$ is of no help so we can remove it). The details are left to the reader.

Remark 3.5 Let $\Omega \subset \mathbf{R}^n$ be a bounded domain with smooth boundary $\partial\Omega$. Consider the Dirichlet problem

$$(3.13) \qquad \begin{cases} -\Delta u = f(x,u) & \text{in } \Omega \\ u = 0 & \text{on } \partial\Omega, \end{cases}$$

where $f \in C(\overline{\Omega} \times \mathbf{R}, \mathbf{R})$. Let $F(x,u) := \int_0^u f(x,s)\,ds$. The functional corresponding to (3.13) is

$$\Phi(u) := \frac{1}{2}\int_\Omega |\nabla u|^2\,dx - \int_\Omega F(x,u)\,dx.$$

If F satisfies asymptotic conditions of the same type as (3.9), (3.10), then $\Phi \in C^1(H_0^1(\Omega), \mathbf{R})$, critical points of Φ correspond to weak solutions of (3.13) [38, 39] and one can obtain results similar to Theorem 3.1. This problem has been considered by several authors, see e.g [5, 37, 42, 43] and the references there.

4 Hamiltonian systems with even Hamiltonian

Suppose that $H \in C(\mathbf{R}^{2N} \times \mathbf{R}, \mathbf{R})$, $H_z \in C(\mathbf{R}^{2N} \times \mathbf{R}, \mathbf{R}^{2N})$, H is 2π-periodic in t and consider the problem

$$(4.1) \qquad \dot{z} = JH_z(z,t), \qquad z(t+2\pi) = z(t).$$

Let $E := H^{1/2}(S^1, \mathbf{R}^{2N})$ be the Sobolev space of 2π-periodic $\mathbf{R}^{2N}$-valued functions

$$(4.2) \qquad z(t) = a_0 + \sum_{k=1}^{\infty} a_k \cos kt + b_k \sin kt, \qquad a_0, a_k, b_k \in \mathbf{R}^{2N}$$

such that $\sum_{k=1}^{\infty} k(|a_k|^2 + |b_k|^2) < \infty$. E is a Hilbert space with an inner product

$$\langle z, z' \rangle := 2\pi a_0 \cdot a_0' + \pi \sum_{k=1}^{\infty} k(a_k \cdot a_k' + b_k \cdot b_k').$$

It is well-known [38, Proposition 6.6] that the space E is compactly embedded in $L^q(S^1, \mathbf{R}^{2N})$ for each $1 \le q < \infty$ (but $z \in E$ does not imply $z \in L^\infty(S^1, \mathbf{R}^{2N})$). Let

$$\Phi(z) := \frac{1}{2}\int_0^{2\pi} (-J\dot{z} \cdot z)\,dt - \int_0^{2\pi} H(z,t)\,dt.$$

If

$$|H_z(z,t)| \le a(1+|z|^s) \tag{4.3}$$

for some $a > 0$ and $s \in [1,\infty)$, then Φ is well-defined on E, $\Phi \in C^1(E,\mathbf{R})$ and $\nabla\Phi(z) = 0$ if and only if $z = z(t)$ is a solution of (4.1) [38, Section 6]. (Note that since $\int_0^{2\pi}(-J\dot{z}\cdot z)\,dt = 2\int_0^{2\pi} p\cdot\dot{q}\,dt$, our functional Φ is the same as the functional I in [38].)

Let A be a symmetric $2N \times 2N$ constant matrix and consider the problem

$$\dot{z} = JAz, \qquad z(t+2\pi) = z(t). \tag{4.4}$$

If $L : E \to E$ is the linear operator given by

$$\langle Lz, y\rangle = \int_0^{2\pi} (-J\dot{z} - Az)\cdot y\,dt, \tag{4.5}$$

then $Lz = 0$ if and only if z satisfies (4.4). Furthermore, using (4.2) we see that

$$\langle Lz, z\rangle = -2\pi A a_0 \cdot a_0 + \pi \sum_{k=1}^{\infty} k[-Jb_k - \frac{1}{k}Aa_k)\cdot a_k + (Ja_k - \frac{1}{k}Ab_k)\cdot b_k]. \tag{4.6}$$

The restriction of this quadratic form to a subspace corresponding to a fixed $k \ge 1$ is represented by the matrix $\pi k T_k(A)$, where

$$T_k(A) := \begin{pmatrix} -\frac{1}{k}A & -J \\ J & -\frac{1}{k}A \end{pmatrix}.$$

Let $M^-(B)$, $M^0(B)$ and $M^+(B)$ denote the number of negative, zero and positive eigenvalues of a symmetric matrix B (counted with their multiplicity). Then $M^0(T_k(A)) = 0$ and $M^\pm(T_k(A)) = 2N$ for all k large enough. Indeed, a simple computation shows that the matrix

$$\begin{pmatrix} 0 & -J \\ J & 0 \end{pmatrix}$$

has the eigenvalues ± 1, each of multiplicity $2N$, so by a perturbation argument, $M^\pm(T_k(A)) = 2N$ for almost all k (cf. [1, Section 12], [2, Section 2]). Therefore the following numbers are well-defined and finite:

$$i^-(A) := M^+(A) - N + \sum_{k=1}^{\infty}(M^-(T_k(A)) - 2N), \tag{4.7}$$

$$i^+(A) := M^-(A) - N + \sum_{k=1}^{\infty}(M^+(T_k(A)) - 2N) \tag{4.8}$$

and

$$i^0(A) := M^0(A) + \sum_{k=1}^{\infty} M^0(T_k(A)). \tag{4.9}$$

Clearly, $i^-(A) + i^+(A) + i^0(A) = 0$. Similar quantities have been introduced by Amann and Zehnder [1, 2] and Benci [9]. Our definition follows that of Li and Liu [28] (to be more precise, the numbers $i^{\pm}(A)$ here and in [28] differ by N). Note that although the Morse index of (4.6) necessarily equals $+\infty$, $i^{-}(A)$ is always finite (and measures in some sense the difference between the negative parts of the quadratic forms $z \mapsto \int_0^{2\pi}(-J\dot{z} - Az)\cdot z\,dt$ and $z \mapsto \int_0^{2\pi}(-J\dot{z}\cdot z)\,dt$). As we shall see, $i^{-}(A)$ will play the same role for Hamiltonian systems as the Morse index does for systems of second order. Let us also remark that in some cases the numbers $i^{\pm}(A)$ can be explicitly computed in terms of purely imaginary eigenvalues of JA [1, Section 12], [2, Section 2].

It is easy to see that $i^0(A) = \dim N(L)$ and $i^0(A)$ is the number of linearly independent solutions of (4.4). Hence in particular $i^0(A) \le 2N$. Moreover, $i^0(A) = 0$ if and only if $\sigma(JA) \cap i\mathbf{Z} = \emptyset$. Indeed, it follows from (4.6) (or by substituting (4.2) into (4.4)) that (4.4) has a nontrivial solution if and only if either A is singular (so $0 \in \sigma(JA)$) or

(4.10)
$$-kJb_k = Aa_k \quad \text{and} \quad kJa_k = Ab_k$$

for some $(a_k, b_k) \ne (0,0)$ and $k \ge 1$. (4.10) is equivalent to

$$JA(a_k - ib_k) = ik(a_k - ib_k).$$

Hence $\pm ik \in \sigma(JA)$ and (4.4) has a nontrivial solution if and only if $\sigma(JA) \cap i\mathbf{Z} \ne \emptyset$.

Let now A_0 and A_∞ be two symmetric constant $2N \times 2N$ matrices. Suppose that H satisfies

(4.11)
$$H(z,t) = \frac{1}{2}A_0 z \cdot z + G_0(z,t),$$

where $(G_0)_z(z,t) = o(|z|)$ uniformly in t as $z \to 0$ and

(4.12)
$$H(z,t) = \frac{1}{2}A_\infty z \cdot z + G_\infty(z,t),$$

where $(G_\infty)_z(z,t) = o(|z|)$ uniformly in t as $|z| \to \infty$.

(4.11) implies that (4.1) has the trivial solution $z = 0$. If H satisfies (4.12), then (4.3) holds with $s = 1$ and therefore $\Phi \in C^1(E, \mathbf{R})$.

Theorem 4.1 *Suppose $H \in C(\mathbf{R}^{2N} \times \mathbf{R}, \mathbf{R})$, $H_z \in C(\mathbf{R}^{2N} \times \mathbf{R}, \mathbf{R}^{2N})$, H is 2π-periodic in t and $H(-z,t) = H(z,t)$ for all z, t. Suppose also H satisfies (4.11), (4.12) and $\sigma(JA_\infty) \cap i\mathbf{Z} = \emptyset$.*
(i) If $i^{-}(A_0) > i^{-}(A_\infty)$, then (4.1) has at least $i^{-}(A_0) - i^{-}(A_\infty)$ pairs of nontrivial solutions.
(ii) If $i^{+}(A_0) > i^{+}(A_\infty)$, then (4.1) has at least $i^{+}(A_0) - i^{+}(A_\infty)$ pairs of nontrivial solutions.

Theorem 4.1 is a slight generalization of [2, Theorem 3] where it was assumed that H_{zz} is continuous and bounded. This assumption was needed in order to reduce the problem to a finite dimensional one. After such reduction Theorem 2.3 was employed. A more general result than Theorem 4.1 has been recently obtained by Fei [23], cf. Remark 4.4 below.

Similarly as in Section 3, we introduce the notation

$$(4.13)\quad \Phi(z) = \frac{1}{2}\int_0^{2\pi} (-J\dot{z} - A_0 z)\cdot z\,dt - \int_0^{2\pi} G_0(z,t)\,dt =: \frac{1}{2}\langle L_0 z, z\rangle - \varphi_0(z)$$
$$= \frac{1}{2}\int_0^{2\pi} (-J\dot{z} - A_\infty z)\cdot z\,dt - \int_0^{2\pi} G_\infty(z,t)\,dt =: \frac{1}{2}\langle L_\infty z, z\rangle - \varphi_\infty(z).$$

Lemma 4.2 *If $H \in C(\mathbf{R}^{2N}\times\mathbf{R},\mathbf{R})$, $H_z \in C(\mathbf{R}^{2N}\times\mathbf{R},\mathbf{R}^{2N})$, H is 2π-periodic in t, $H(0,t)=0$ for all t and (4.11), (4.12) are satisfied, then $\nabla\varphi_0(z) = o(\|z\|)$, $\varphi_0(z) = o(\|z\|^2)$ as $z\to 0$ and $\nabla\varphi_\infty(z) = o(\|z\|)$, $\varphi_\infty(z) = o(\|z\|^2)$ as $\|z\|\to\infty$.*

Proof Note first that $H(0,t)=0$ implies $\varphi_0(0)=\varphi_\infty(0)=0$. For each $\varepsilon>0$ there is a $C_1(\varepsilon)$ such that $|(G_0)_z(z,t)| \le \varepsilon|z| + C_1(\varepsilon)|z|^2$. Hence

$$|\langle\nabla\varphi_0(z), y\rangle| \le \int_0^{2\pi} (\varepsilon|z| + C_1(\varepsilon)|z|^2)|y|\,dt \le (\varepsilon\|z\| + C_2(\varepsilon)\|z\|^2)\|y\|,$$

where we have used that $E \hookrightarrow L^q(S^1,\mathbf{R}^{2N})$ for $q=2$ and 3. Taking the supremum over $\|y\|\le 1$, dividing by $\|z\|$ and letting $z\to 0$ we see that $\nabla\varphi_0(z) = o(\|z\|)$ as $z\to 0$. Since

$$\varphi_0(z) = \int_0^1 \frac{d}{ds}\varphi_0(sz)\,ds = \int_0^1 \langle\nabla\varphi_0(sz), z\rangle\,ds,$$

$\varphi_0(z) = o(\|z\|^2)$.

Similarly, for each $\varepsilon>0$ there is a $C_3(\varepsilon)$ such that $|(G_\infty)_z(z,t)| \le \varepsilon|z| + C_3(\varepsilon)$. So

$$(4.14)\qquad |\langle\nabla\varphi_\infty(z), y\rangle| \le (\varepsilon\|z\| + C_4(\varepsilon))\|y\|$$

and $\nabla\varphi_\infty(z) = o(\|z\|)$ as $\|z\|\to\infty$. Moreover, according to (4.14),

$$|\varphi_\infty(z)| = \left|\int_0^1 \langle\nabla\varphi_\infty(sz), z\rangle\,ds\right| \le \frac{1}{2}\varepsilon\|z\|^2 + C_4(\varepsilon)\|z\|.$$

Hence $\varphi_\infty(z) = o(\|z\|^2)$ as $\|z\|\to\infty$. □

Let now

$$E_n := \{z\in E : z(t) = a_0 + \sum_{k=1}^{n} a_k\cos kt + b_k + \sin kt\},$$

$d_n := \frac{1}{2}\dim E_n = N(1+2n)$ and $\mathcal{E} := \{E_n, d_n\}_{n=1}^{\infty}$. The operators L_0 and L_∞ map each E_n into itself.

Lemma 4.3 *If $H \in C(\mathbf{R}^{2N}\times\mathbf{R},\mathbf{R})$, $H_z \in C(\mathbf{R}^{2N}\times\mathbf{R},\mathbf{R}^{2N})$, H is 2π-periodic in t, satisfies (4.12) and $\sigma(JA_\infty)\cap i\mathbf{Z} = \emptyset$, then Φ satisfies (PS) and (PS)*.*

Proof The argument of Lemma 3.3 can be applied here. Observe only that if $z_j \to \overline{z}$ weakly in E, then $z_j \to \overline{z}$ strongly in $L^2(S^1, \mathbf{R}^{2N})$, and since (4.3) is satisfied with $s = 1$, $P_{n_j}\nabla\varphi_\infty(z_j) \to \nabla\varphi_\infty(\overline{z})$ (cf. [38, Appendix B]). □

Proof of Theorem 4.1 As in the proof of Theorem 3.1, we may assume that $H(0,t) = 0$ for all t. So $\Phi(0) = 0$. Moreover, Φ is an even functional.

(i) Let $l := i^-(A_0)$, $m := i^-(A_\infty)$, $X := E^+(L_\infty)$ and $Y := E^-(L_0)$. Using Lemma 4.2 we see as in the proof of Theorem 3.1 that $\Phi|_X \geq \alpha$ and $\Phi|_{Y\cap\partial B_\rho} \leq \beta$ for a suitable choice of $\alpha < \beta < 0$ and $\rho > 0$.

If n is large enough, then $M^\pm(T_k(A)) = 2N$ for $k > n$, and we obtain from (4.6), (4.7) that

$$\begin{aligned}
\operatorname{codim}_{E_n} X_n &= \operatorname{codim}_{E_n} E^+(L_\infty) \cap E_n = \dim E^-(L_\infty) \cap E_n \\
&= M^+(A_\infty) + \sum_{k=1}^{n} M^-(T_k(A)) \\
&= M^+(A_\infty) - N + \sum_{k=1}^{n} (M^-(T_k(A)) - 2N) + N(1 + 2n) \\
&= m + d_n.
\end{aligned}$$

Similarly,

$$\dim Y_n = \dim E^-(L_0) \cap E_n = l + d_n.$$

So Φ satisfies the hypotheses of Theorem 2.8.

(ii) Take $l := i^+(A_0)$, $m := i^+(A_\infty)$, $X := E^-(L_\infty)$, $Y := E^+(L_0)$ and apply the same argument to $-\Phi$. □

Remark 4.4 (i) If A is a nonconstant matrix with 2π-periodic entries, the definitions of $i^\pm(A)$ and $i^0(A)$ no longer make sense. Therefore we need some other quantities to measure the size of the positive and the negative part of L. Assume for simplicity that the operator L corresponding to A via (4.5) is invertible and let M_n^- be the Morse index of the quadratic form $z \mapsto \langle Lz, z\rangle$ restricted to E_n. Let

$$j^-(A) := \lim_{n\to\infty} (M_n^- - d_n)$$

and $j^+(A) := j^-(-A)$. It can then be shown (after some work) that $j^\pm(A)$ are well-defined and finite, and $j^\pm(A) = i^\pm(A)$ if A is a constant matrix. See [26, Sections 7 and 5], where $j^-(A)$ is called the $\mathcal{E}$-Morse index and is denoted by $M_{\mathcal{E}}^-(L)$. Moreover, $j^-(A)$ equals the Maslov index corresponding to A, cf. [13, Theorem IV.1.2]. More information about the Maslov index may be found e.g. in [13, 18, 31, 32]. Since Φ satisfies (PS)* [26, Lemma 7.3], it is easy to see that the conclusions of Theorem 4.1 remain valid if L_0 and L_∞ are invertible.

If L is not invertible, the definition of $j^-(A)$ needs to be modified, see [26, Section 5].

A different approach, also allowing A_0 and A_∞ to be nonconstant 2π-periodic, has been worked out by Fei [23]. In [23] a certain relative Morse index of the pair A_0, A_∞ is introduced and is shown to be equal to the difference between the Maslov indices of A_0 and A_∞. The conclusion of Theorem 4.1 is then obtained by applying Benci's theorem [9] mentioned in Section 2.

(ii) Conclusions similar to those contained in (ii) of Remark 3.4 are also valid here. Since E is not a subspace of the space of continuous functions, the situation is now more difficult. Still it can be shown that Φ satisfies (PS)* if $(G_\infty)_z$ is bounded and $G_\infty(z,t) \to \infty$ (or $-\infty$) uniformly in t as $|z| \to \infty$ [26, Lemma 7.3] and that $\varphi_0(z) > 0$ in a deleted neighbourhood of $0 \in E$ if $G_0(z,t) > 0$ for $0 < |z| < \delta$ [29, Lemma 2.2], see also [28, Lemma 3.4]. The computations of $\dim X_n$ and $\dim Y_n$ under different sets of hypotheses may be found in [26, Section 7].

Remark 4.5 Let $\Omega \subset \mathbf{R}^n$ be a bounded domain with smooth boundary, let $F, F_u, F_v \in C(\overline{\Omega} \times \mathbf{R}^2, \mathbf{R})$ and consider the Dirichlet problem

$$(4.15) \qquad \begin{cases} -\Delta u = F_v(x,u,v) & \text{in } \Omega \\ -\Delta v = F_u(x,u,v) & \text{in } \Omega \\ u = v = 0 & \text{on } \partial\Omega. \end{cases}$$

The functional corresponding to (4.15) is

$$\Phi(u,v) = \int_\Omega \nabla u \cdot \nabla v \, dx - \int_\Omega F(x,u,v)\, dx.$$

If F satisfies conditions similar to (4.11), (4.12), then $\Phi \in C^1(H_0^1(\Omega) \times H_0^1(\Omega), \mathbf{R})$ and critical points of Φ are weak solutions of (4.15). It is easy to see that Φ is strongly indefinite and the quadratic form corresponding to the linearization of (4.15) has infinite Morse index. However, also here one can introduce finite indices like $i^\pm(A)$ and $i^0(A)$ ([26, Section 9], cf. also [41, Section 5]). Having these indices it is easy to obtain results similar to Theorem 4.1. The details are left to the reader.

Remark 4.6 If V and H are not assumed to be even and other hypotheses of Theorems 3.1 and 4.1 are satisfied, then it can be shown that under some additional technical hypotheses (3.1) and (4.1) have a nontrivial solution (or two nontrivial solutions if the difference between the indices is large enough). For (3.1) these results are obtained by Morse theory, see [34] and the references there. Problem (4.1) can be approached either by performing a finite dimensional reduction and then applying the usual Morse theory to the reduced functional [1, 2, 11, 13, 18, 31] or by working directly in the infinite dimensional space E and using a suitable generalized Morse theory [26, 28, 40].

Let us also mention that there is a vast literature concerning problem (3.13) with F satisfying our asymptotic conditions but not necessarily being even (cf. [13, Chapter III]). Results on (4.15) under such hypotheses may be found in [26].

5 Index and minimax principles for S^1-invariant functionals

Let E be a Banach space and $\mathbf{T} = \{T_\theta\}_{\theta\in S^1}$ an isometric representation of S^1 in E. Let Σ be the family of closed S^1-invariant subsets of E, i.e.,

$$\Sigma := \{A \in E : A \text{ is closed and } T_\theta A = A \text{ for each } \theta\}$$

(more precisely, the sets in Σ are invariant with respect to the representation $\mathbf{T}$). Recall that if $\mathbf{T} = \{T_\theta\}$ and $\mathbf{S} = \{S_\theta\}$ are isometric representations of S^1 in respectively E_1 and E_2, then a mapping $f : E_1 \to E_2$ is called equivariant provided $f(T_\theta x) = S_\theta f(x)$ for all x, θ. If $f : E_1 \to E_2$, then the mapping $f_G : E_1 \to E_2$ given by

$$f_G(x) := \frac{1}{2\pi}\int_0^{2\pi} S_{-\theta} f(T_\theta x)\, d\theta \tag{5.1}$$

is easily seen to be equivariant (although $G = S^1$ here, we prefer to use the notation f_G instead of f_{S^1}; similarly, we will write E^G rather than E^{S^1}).

Definition 5.1 Let $A \in \Sigma$, $A \neq \emptyset$. The S^1-*index* of A, denoted $i(A)$, is the smallest integer k for which there exists a mapping $f \in C(A, \mathbf{C}^k - \{0\})$ such that $f(T_\theta x) = e^{in\theta} f(x)$ for some $n \in \mathbf{N} - \{0\}$ and all x, θ. If such a mapping does not exist for any k, then $i(A) := +\infty$. Moreover, we set $i(\emptyset) := 0$.

The above definition is due to Benci [8, 9]. Properties of this index may be found e.g. in [8, 34]. We do not formulate them here because below we will consider a more general situation. Let us also remark that a different, cohomological S^1-index, has been introduced by Fadell and Rabinowitz [22]. While Benci's index is much more elementary, the index of Fadell and Rabinowitz has certain additional properties which are sometimes useful (but will not be needed here).

Remark 5.2 By the Peter-Weyl theorem, if $\mathbf{S}$ is an isometric representation of S^1 in $\mathbf{R}^m$, then there exists an orthonormal basis such that $\mathbf{R}^m = \mathbf{R}^s \times \mathbf{R}^{2k} \cong \mathbf{R}^s \times \mathbf{C}^k$ and

$$S_\theta(x_1, \ldots, x_s, z_1, \ldots, z_k) = (x_1, \ldots, x_s, e^{in_1\theta} z_1, \ldots, e^{in_k\theta} z_k),$$

where $n_1, \ldots, n_k \in \mathbf{N} - \{0\}$, see e.g. [34, Theorem 5.1]. Note that the fixed point set $E^G = \mathbf{R}^s \times \{0\}$. It follows that in Definition 5.1 we may equivalently require that there exists a mapping $f \in C(A, \mathbf{C}^k - \{0\})$ and a representation $\mathbf{S}$ of S^1 in $\mathbf{C}^k$ such that $f \circ T_\theta = S_\theta \circ f$ for each θ and $E^G(\mathbf{S}) = \{0\}$ ($E^G(\mathbf{S})$ denotes the fixed point set of the action represented by $\mathbf{S}$). Indeed, if $f \in C(A, \mathbf{C}^k - \{0\})$, then we may assume after a change of coordinates that $f = (f_1, \ldots, f_k)$ and $f_j(T_\theta x) = e^{in_j\theta} f_j(x)$, $n_j \in \mathbf{N} - \{0\}$, $j = 1, \ldots, k$. Let $g_j(x) = f_j(x)^{n/n_j}$, where n is the least common multiple of $n_1, \ldots, n_k$. Then $g : A \to \mathbf{C}^k - \{0\}$ and $g(T_\theta x) = e^{in\theta} g(x)$.

Remark 5.3 If $A \in \Sigma$ and $A \cap E^G \neq \emptyset$, then $i(A) = \infty$. Indeed, let f be as in Definition 5.1. For $x \in A \cap E^G$ and each θ we have $f(x) = f(T_\theta x) = e^{in\theta} f(x)$. Hence $f(x) = 0$, a contradiction.

In what follows we always assume that $\mathbf{T}$ is a fixed isometric representation of S^1 in E. So the family Σ is invariant with respect to this $\mathbf{T}$ and $f : E \to E$ is equivariant whenever $f(T_\theta x) = T_\theta f(x)$ for all x, θ. Suppose

(5.2) $\qquad F$ is a finite dimensional subspace of E^G and $E^G = F \oplus \widetilde{F}$.

Definition 5.4 Let $A \in \Sigma$, $A \neq \emptyset$. The S^1-*index* of A *relative* to F, denoted $i_F(A)$, is the smallest integer k for which there exists a mapping $f = (f_1, f_2) \in C(A, E^G \times \mathbf{C}^k - \widetilde{F} \times \{0\})$ such that $f(T_\theta x) = (f_1(x), e^{in\theta} f_2(x))$ for some $n \in \mathbf{N} - \{0\}$ and all x, θ, and there is a homotopy joining $f_1|_{A \cap E^G}$ to $id_{A \cap E^G}$ in $E^G - \widetilde{F}$. We also set $i_F(A) := +\infty$ if such a mapping does not exist for any k, and $i_F(\emptyset) := 0$.

Observe that the choice of both F and $\widetilde{F}$ is essential in the definition of $i_F(A)$. So the notation $i_{F,\tilde{F}}(A)$ would be more accurate; however, for the sake of convenience we prefer to write $i_F(A)$.

Remark 5.5 Equivalently one can define $i_F(A)$ to be the smallest integer k for which there exists a mapping $f = (f_1, f_2) \in C(A, F \times \mathbf{C}^k - \{(0,0)\})$ such that $f(T_\theta x) = (f_1(x), e^{in\theta} f_2(x))$ for some $n \in \mathbf{N} - \{0\}$ and all x, θ, and there is a homotopy joining $f_1|_{A \cap E^G}$ to $P_F|_{A \cap E^G}$ in $F - \{0\}$, where P_F is the projector onto F associated with the decomposition (5.2).

Remark 5.6 Let $\eta_1 : (A \cap E^G) \times [0,1] \to E^G - \widetilde{F}$ be the homotopy joining $f_1|_{A \cap E^G}$ to $id_{A \cap E^G}$. Write $\eta = (\eta_1, \eta_2) : (A \cap E^G) \times [0,1] \to E^G \times \mathbf{C}^k - \widetilde{F} \times \{0\}$, where $\eta_2 \equiv 0$. By the homotopy extension theorem [12, Theorems 8.1 and 10.1], η can be extended to a mapping $\tilde{\eta} = (\tilde{\eta}_1, \tilde{\eta}_2) : A \times [0,1] \to E^G \times \mathbf{C}^k - \widetilde{F} \times \{0\}$ in such a way that $\tilde{\eta}(x, 0) = f(x)$ (below we sketch a proof). Using (5.1) we may assume $\tilde{\eta}(T_\theta x, t) = (\tilde{\eta}_1(x,t), e^{in\theta} \tilde{\eta}_2(x,t))$. Set $\tilde{f}(x) := \tilde{\eta}(x, 1)$. Then $\tilde{f}_1|_{A \cap E^G} = id_{A \cap E^G}$. Therefore the condition in Definition 5.4 that $f_1|_{A \cap E^G}$ is homotopic to $id_{A \cap E^G}$ may be replaced by the seemingly stronger one that $f_1|_{A \cap E^G} = id_{A \cap E^G}$. Similarly, in Remark 5.5 the condition that $f_1|_{A \cap E^G}$ is homotopic to $P_F|_{A \cap E^G}$ may be replaced by $f_1|_{A \cap E^G} = P_F|_{A \cap E^G}$.

To prove that an extension $\tilde{\eta}$ of the homotopy η exists, let $C := A \times \{0\} \cup (A \cap E^G) \times [0,1]$ and set $\eta(x, 0) = f(x)$ for $x \in A$. Then $\eta : C \to E^G \times \mathbf{C}^k - \widetilde{F} \times \{0\}$. Extend η to a mapping $\hat{\eta} : A \times [0,1] \to E^G \times \mathbf{C}^k$ and let $V := \hat{\eta}^{-1}(E^G \times \mathbf{C}^k - \widetilde{F} \times \{0\})$. Then V is an open neighbourhood of C in $A \times [0,1]$. There exists an open neighbourhood U of $A \cap E^G$ in A such that $(A \cap E^G) \times [0,1] \subset V$ (indeed, for each $x \in A \cap E^G$ we can find an open neighbourhood U_x of x in A with $U_x \times [0,1] \subset V$; now take $U := \bigcup_{x \in A \cap E^G} U_x$). Let $\alpha \in C(A, [0,1])$ be a function satisfying $\alpha(x) = 0$ for $x \in A - U$ and $\alpha(x) = 1$ for $x \in A \cap E^G$. Then $\tilde{\eta}(x,t) := \hat{\eta}(x, \alpha(x)t)$ has the required properties.

Note that if $A \in \Sigma$ and $A \subset E^G$, then $i_F(A) = 0$ whenever $A \cap \widetilde{F} = \emptyset$ and $i_F(A) = +\infty$ otherwise. Note also that if $F = \{0\}$, then $i_F \equiv i$. If $F = E^G$, then in view of Remark 5.6, i_F coincides with the relative index γ_0 introduced by Berestycki et al. in [10]. Let us mention here that there also exist different *cohomological* relative index theories, see e.g. [20, 21, 22], and in particular [6], where an up-to-date list of references may be found.

If X is an invariant subspace of E and X admits a topological complement, then it also admits an invariant topological complement. Indeed, if P is a continuous projector of E onto X, then the mapping P_G given by

$$P_G x := \frac{1}{2\pi}\int_0^{2\pi} T_{-\theta} P T_\theta x \, d\theta$$

is easily seen to be an equivariant projector onto X and the nullspace of P_G is an invariant complement of X (cf. [34, Lemma 5.1]).

Since the mapping Q given by

$$Qx := \frac{1}{2\pi}\int_0^{2\pi} T_\theta x \, d\theta$$

is a continuous projector onto E^G, there exists an invariant subspace $\widetilde{E}$ such that

$$E = E^G \oplus \widetilde{E}. \tag{5.3}$$

Proposition 5.7 *Let $A, B \in \Sigma$ and let $F, \widetilde{F}, \widetilde{E}$ be as in (5.2), (5.3).*
(i) If there exists an equivariant mapping $g \in C(A, B)$ such that $g|_{A \cap E^G}$ is homotopic to $id_{A \cap E^G}$ in $E^G - \widetilde{F}$, then $i_F(A) \le i_F(B)$.
(ii) $i_F(A \cup B) \le i_F(A) + i(B)$.
(iii) If A is compact, then there exists a neighbourhood $N \in \Sigma$ of A such that $i(A) = i(N)$.
(iv) Let $R > 0$. If $Y \subset \widetilde{E}$ is an invariant subspace of dimension $2l$, then $i_F((F \oplus Y) \cap \partial B_R) = l$.
(v) If X is an invariant subspace of $\widetilde{E}$ such that $\operatorname{codim}_{\widetilde{E}} X = 2m$ *and if $i_F(A) > m$, then $A \cap (\widetilde{F} \oplus X) \neq \emptyset$.*

Proof (i) If $i_F(B) = +\infty$, the conclusion is trivially satisfied. So assume $i_F(B) = k < \infty$. Let $f \in C(B, E^G \times \mathbf{C}^k - \widetilde{F} \times \{0\})$ have the properties specified in Definition 5.4. Then $f \circ g \in C(A, E^G \times \mathbf{C}^k - \widetilde{F} \times \{0\})$ and $(f \circ g)(T_\theta x) = ((f_1 \circ g)(x), e^{in\theta}(f_2 \circ g)(x))$. Let $\eta : (B \cap E^G) \times [0,1] \to E^G - \widetilde{F}$ and $\gamma : (A \cap E^G) \times [0,1] \to E^G - \widetilde{F}$ be the homotopies joining respectively $f_1|_{B \cap E^G}$ to $id_{B \cap E^G}$ and $g|_{A \cap E^G}$ to $id_{A \cap E^G}$. Then

$$\delta(x,t) := \begin{cases} \eta(g(x), 2t), & 0 \le t \le \frac{1}{2} \\ \gamma(x, 2t-1), & \frac{1}{2} \le t \le 1 \end{cases}$$

is a homotopy joining $(f_1 \circ g)|_{A \cap E^G}$ to $id_{A \cap E^G}$ in $E^G - \widetilde{F}$. Hence $i_F(A) \le k$.

(ii) Set $i_F(A) = k$, $i(B) = m$, where we may assume $k, m < \infty$. Let $f = (f_1, f_2) \in C(A, E^G \times \mathbf{C}^k - \widetilde{F} \times \{0\})$ be as in Definition 5.4 and $g \in C(B, \mathbf{C}^m - \{0\})$ as in Definition 5.1. Then in particular

(5.4) $\quad f_2(T_\theta x) = e^{in\theta} f_2(x) \quad$ and $\quad g(T_\theta x) = e^{ip\theta} g(x) \quad$ for some $n, p \in \mathbf{N} - \{0\}$.

Using (5.1) with respectively $S_\theta = e^{in\theta}$ and $S_\theta = e^{ip\theta}$ we may extend f and g in such a way that $f \in C(E, E^G \times \mathbf{C}^k)$, $g \in C(E, \mathbf{C}^m)$ and (5.4) is satisfied. Let

$$h(x) := (f_1(x), f_2(x)^p, g(x)^n) \equiv (f_1(x), h_2(x)).$$

Then $h(T_\theta x) = (f_1(x), e^{inp\theta} h_2(x))$ and $h|_{A \cup B} \in C(A \cup B, E^G \times \mathbf{C}^{k+m} - \widetilde{F} \times \{0\})$. Since $i(B) < \infty$, $B \cap E^G = \emptyset$ and $(A \cup B) \cap E^G = A \cap E^G$. Hence $f_1|_{(A \cup B) \cap E^G} = f_1|_{A \cap E^G}$ and there is a homotopy joining $f_1|_{(A \cup B) \cap E^G}$ to $id_{(A \cup B) \cap E^G}$ in $E^G - \widetilde{F}$.

(iii) Assume $i(A) = k < \infty$. There exists a mapping $f \in C(A, \mathbf{C}^k - \{0\})$ with the properties stated in Definition 5.1. As in (ii), we may extend f in such a way that $f \in C(E, \mathbf{C}^k)$ and $f(T_\theta x) = e^{in\theta} f(x)$. There is a closed neighbourhood $N \in \Sigma$ of A such that $f|_N \in C(N, \mathbf{C}^k - \{0\})$. Hence $i(N) \leq k$. Since $i(N) \geq k$ (take $g : A \hookrightarrow N$ and employ (i) with $F = \{0\}$), we obtain the conclusion.

(iv) Since Y may be identified with $\mathbf{C}^l$, it follows that $i_F((F \oplus Y) \cap \partial B_R) \leq l$. The fact that there is no mapping $f \in C((F \oplus Y) \cap \partial B_R, F \times \mathbf{C}^{l-1} - \{(0,0)\})$ satisfying the properties given in Remark 5.5 is an S^1-version of the Borsuk-Ulam theorem. The proof in the case $F = \{0\}$ may be found in [34, Theorem 5.5]. For the general case, see Marzantowicz [33] or Nirenberg [35]. To be more precise, we may assume via Remark 5.5 that $f \in C((F \times \mathbf{C}^l) \cap \partial B_R, F \times \mathbf{C}^l - \{(0,0)\})$. Since the range of f is in $F \times \mathbf{C}^{l-1}$, the Brouwer degree $\deg(f, (F \times \mathbf{C}^l) \cap B_R, 0) = 0$. On the other hand, since $f_1|_{F \cap \partial B_R}$ is homotopic to $P_F|_{F \cap \partial B_R} = id_{F \cap \partial B_R}$ in $F - \{0\}$, it follows that $\deg(f_1, F \cap B_R, 0) = 1$ and, according to [33, Corollary 2.8], $\deg(f, (F \times \mathbf{C}^l) \cap B_R, 0) \neq 0$, a contradiction. Another proof that the degree is nonzero - which is more elementary than in [33] - may be found in [35, Theorem 3]. Note that here $f_1|_{F \cap \partial B_R}$ is homotopic to $id_{F \cap \partial B_R}$ while in [35] $f_1|_{F \cap \partial B_R} = id_{F \cap \partial B_R}$. However, in view of Remark 5.6, these two conditions are equivalent.

(v) There exists an invariant subspace Y of $\widetilde{E}$ such that $\dim Y = 2m$ and $\widetilde{E} = X \oplus Y$. Suppose $A \cap (\widetilde{F} \oplus X) = \emptyset$ and let $f(x) := P_F x + P_Y x$, where P_F, P_Y are the projectors onto F and Y corresponding to the decomposition $E = F \oplus \widetilde{F} \oplus X \oplus Y$. Then $f \in C(A, F \oplus Y - \{0\})$, f is equivariant and $f(x) = P_F x$ for $x \in A \cap E^G$. Identifying $F \oplus Y$ with $F \times \mathbf{C}^m$ we obtain $f = (f_1, f_2) \in C(A, F \times \mathbf{C}^m - \{(0,0)\})$, $f_1|_{A \cap E^G} = P_F|_{A \cap E^G}$ and according to Remark 5.2 we may modify f_2 so that $f_2(T_\theta x) = e^{in\theta} f_2(x)$. Hence $i_F(A) \leq m$, a contradiction. □

Recall that the orbit of x is the set $\mathcal{O}(x) = \{T_\theta x : \theta \in S^1\}$.

Proposition 5.8 *If $A \in \Sigma$, $A \cap E^G = \emptyset$ and $i(A) \geq 2$, then A contains infinitely many distinct orbits.*

Proof Assume A consists of finitely many orbits $\mathcal{O}(x_1), \ldots, \mathcal{O}(x_k)$. Clearly, $T_{2\pi}x_j = x_j$. Since $x_j \notin E^G$, there exists a largest integer $n_j \geq 1$ such that $T_{2\pi/n_j}x_j = x_j$. Let n be the least common multiple of $n_1, \ldots, n_k$ and set

$$f(T_\theta x_j) := e^{in\theta}, \qquad 1 \leq j \leq k.$$

Then $f \in C(A, \mathbf{C} - \{0\})$ and therefore $i(A) = 1$. □

Let $\mathcal{E} = \{E_n, d_n\}_{n=1}^{\infty}$, where E_n and d_n are as in Section 2 and in addition all E_n are S^1-invariant and $E^G \subset E_n$ for some n. Recall the notation $A_n := A \cap E_n$.

Definition 5.9 Let $A \in \Sigma$. The *relative S^1-limit index* of A *with respect* to $\mathcal{E}$, $i_{F,\mathcal{E}}(A)$, is defined by

$$i_{F,\mathcal{E}}(A) := \limsup_{n\to\infty}(i_F(A_n) - d_n).$$

Also in the S^1-context the notion of (non-relative) limit index is due to Li [30]. If $E_n = E$ and $d_n = 0$ for all n, then $i_{F,\mathcal{E}} \equiv i_F$.

Proposition 5.10 *Let $A, B \in \Sigma$ and let $F, \widetilde{F}, \widetilde{E}$ be as in (5.2), (5.3).*
(i) If for almost all n there exists an equivariant mapping $g_n \in C(A_n, B_n)$ such that $g_n|_{A\cap E^G}$ is homotopic to $id_{A\cap E^G}$ in $E^G - \widetilde{F}$, then $i_{F,\mathcal{E}}(A) \leq i_{F,\mathcal{E}}(B)$.
(ii) $i_{F,\mathcal{E}}(A \cup B) \leq i_{F,\mathcal{E}}(A) + i(B)$.
(iii) Let $l \in \mathbf{Z}$ and $R > 0$. If Y is an invariant subspace of $\widetilde{E}$ such that $\dim Y_n = 2(d_n + l)$ for almost all n, then $i_{F,\mathcal{E}}((F \oplus Y) \cap \partial B_R) = l$.
(iv) Let $m \in \mathbf{Z}$. If X is an invariant subspace of $\widetilde{E}$ such that $\operatorname{codim}_{\widetilde{E}_n} X_n = 2(d_n + m)$ for almost all n and if $i_{F,\mathcal{E}}(A) > m$, then $A \cap (\widetilde{F} \oplus X) \neq \emptyset$.

Proof Use Definition 5.9, Proposition 5.7 and compare with the proof of Proposition 2.6. Note that in part (i) we have $A \cap E^G = A_n \cap E^G$ for almost all n (because $E^G \subset E_n$). □

Suppose that $\Phi \in C^1(E, \mathbf{R})$ is S^1-invariant. It follows from (5.1) (with $S_\theta = T_\theta$) that there exists an S^1-equivariant pseudo-gradient vector field for Φ. Therefore, if the set N in Lemma 2.4 is S^1-invariant, then the deformation η is equivariant with respect to x, i.e., $\eta(T_\theta x, t) = T_\theta \eta(x, t)$ for all x, t, θ. See [34, Section 6.1] for more details.

The main result of this section is the following

Theorem 5.11 *Suppose that $\Phi \in C^1(E, \mathbf{R})$ is S^1-invariant, satisfies $(PS)^*$, $\Phi(0) = 0$ and $F, \widetilde{F}, \widetilde{E}$ are subspaces satisfying (5.2), (5.3). Moreover, suppose there exist numbers $\rho > 0$, $\alpha < \beta < 0$ and invariant subspaces $X, Y \subset \widetilde{E}$ such that:*
(i) $E^G \subset E_n$ for almost all n and $\{x \in E : \Phi(x) < 0,\ \Phi'(x) = 0\} \cap E^G = \emptyset$;
(ii) $\operatorname{codim}_{\widetilde{E}_n} X_n = 2(d_n + m)$ and $\dim Y_n = 2(d_n + l)$ for almost all n and some $m, l \in \mathbf{Z}$,

$m < l$;
(iii) $\Phi|_{(F\oplus Y)\cap\partial B_\rho} \leq \beta$;
(iv) $\Phi|_{\widetilde{F}\oplus X} \geq \alpha$ and $\Phi|_{\widetilde{F}} \geq 0$.
Then Φ has at least $l-m$ distinct critical orbits $\mathcal{O}(x_j)$ such that $\mathcal{O}(x_j)\cap E^G = \emptyset$. The corresponding critical values can be characterized as

$$c_j = \inf_{i_{F,\varepsilon}(A)\geq j} \sup_{x\in A} \Phi(x), \qquad m+1 \leq j \leq l,$$

and are contained in the interval $[\alpha, \beta]$.

Proof We need to slightly modify the proof of Theorem 2.8. Since $\dim Y_n = 2(d_n + l)$ for almost all n, $i_{F,\varepsilon}((F\oplus Y)\cap\partial B_\rho) = l$ and $c_l \leq \beta$. If $i_{F,\varepsilon}(A) \geq m+1$, then $A\cap(\widetilde{F}\oplus X) \neq \emptyset$. Therefore $c_{m+1} \geq \alpha$. Since $c_j \leq \beta < 0$ for all j, it follows from (i) that $K_{c_j}\cap E^G = \emptyset$. So according to Proposition 5.8, K_{c_j} contains infinitely many distinct orbits whenever $i(K_{c_j}) \geq 2$.

The remaining part of the proof is the same as that of Theorem 2.8 if we can show

$$i_{F,\varepsilon}(\Phi^{c+\varepsilon}) \leq i_{F,\varepsilon}(\Phi^{c-\varepsilon}) + i(N),$$

where N is an equivariant neighbourhood of K_c and $i(N) = i(K_c)$ (such N exists according to (iii) of Proposition 5.7). As in Theorem 2.8 we see from (PS)* that Φ satisfies (2.5) and for each large n there is an equivariant deformation η_n such that

$$\eta_n(\Phi_n^{c+\varepsilon}, 1) \subset \Phi_n^{c-\varepsilon} \cup N_n.$$

Moreover, since $\Phi_n|_{\widetilde{F}} \geq 0$ and we may assume $c+\varepsilon < 0$, η_n is a homotopy joining $id_{\Phi^{c+\varepsilon}\cap E^G}$ to $\eta_n(.,1)|_{\Phi^{c+\varepsilon}\cap E^G}$ in $E^G - \widetilde{F}$. Therefore, by Proposition 5.10,

$$i_{F,\varepsilon}(\Phi^{c+\varepsilon}) \leq i_{F,\varepsilon}(\Phi^{c-\varepsilon}\cup N) \leq i_{F,\varepsilon}(\Phi^{c-\varepsilon}) + i(N). \qquad \square$$

Theorem 5.11 is a (somewhat more general) variant of a result in [30] and is also similar to a result in [9] and [19] (in [9, 30] one has $F = \{0\}$ or $F = E^G$).

Note that if $E = E_n$ and $d_n = 0$ for all n, then Theorem 5.11 is an S^1-analogue of Theorem 2.3. Note also the crucial role played by the assumption (i) from which it follows that $K_{c_j}\cap E^G = \emptyset$. If $K_{c_j}\cap E^G \neq \emptyset$ for some j, then $i(K_{c_j}) = +\infty$ and the argument of Theorem 5.11 may give only one critical point. The only result we know of where the existence of multiple critical orbits in $E - E^G$ has been established without excluding that K_{c_j} intersects E^G is due to Bartsch and Wang [7]. However, the problem considered there is rather different: the symmetry group is $\mathbf{Z}_2$, it acts on a manifold and the fixed point set of the action is finite (but contains more than one point).

6 Autonomous systems

In this section we will again be concerned with the second order system of Section 3 and the Hamiltonian system of Section 4. However, this time we assume that the systems are autonomous, i.e. the potential V and the Hamiltonian H are independent of t (we do not assume V and H are even). Since t does not appear explicitly in the equations, it makes sense to look for periodic solutions of any period T.

Consider the problem

$$(6.1) \qquad -\ddot{x} = V'(x), \qquad x(t+T) = x(t).$$

Substituting $\tau := t/\lambda$, where $\lambda := T/2\pi$, gives

$$(6.2) \qquad -\ddot{x} = \lambda^2 V'(x), \qquad x(\tau + 2\pi) = x(\tau).$$

Hence we may restrict our attention to the existence of 2π-periodic solutions and to the problem

$$(6.3) \qquad -\ddot{x} = V'(x), \qquad x(t+2\pi) = x(t).$$

Let $E := H^1(S^1, \mathbf{R}^N)$,

$$\Phi(x) := \frac{1}{2}\int_0^{2\pi} |\dot{x}|^2\,dt - \int_0^{2\pi} V(x)\,dt$$

and $(T_\theta x)(t) := x(t+\theta)$. Then $\mathbf{T} = \{T_\theta\}$ is an isometric representation of S^1 in E and it is easy to see that $\Phi(T_\theta x) = \Phi(x)$ for all x, θ. Moreover, E^G is the subspace of constant functions, i.e.,

$$E^G = \{x \in E : x(t) = a_0\} \cong \mathbf{R}^N$$

(cf. (3.2)) and

$$E = E^G \oplus \widetilde{E},$$

where $\widetilde{E}$ is the subspace of functions having mean value zero. Clearly, $\widetilde{E}$ is S^1-invariant.

For a symmetric $N \times N$ matrix A with eigenvalues $\mu_1, \ldots, \mu_N$, let

$$(6.4) \qquad i_0^-(A) := i^-(A) - \#\{j : \mu_j > 0\} = 2\sum_{j=1}^N \#\{k \geq 1 : \mu_j > k^2\}$$

and

$$(6.5) \qquad i_0^0(A) := i^0(A) - \#\{j : \mu_j = 0\} = 2\sum_{j=1}^N \#\{k \geq 1 : \mu_j = k^2\}$$

(cf. (3.7), (3.8)). Then $i_0^-(A)$ and $i_0^0(A)$ are respectively the Morse index and the nullity of the restriction of the quadratic form (3.6) to $\widetilde{E}$. The asymptotic conditions (3.9), (3.10) now take the form

$$(6.6) \qquad V(x) = \frac{1}{2}A_0 x \cdot x + W_0(x), \text{ where } W_0'(x) = o(|x|) \text{ as } x \to 0$$

and

$$(6.7) \qquad V(x) = \frac{1}{2}A_\infty x \cdot x + W_\infty(x), \text{ where } W'_\infty(x) = o(|x|) \text{ as } |x| \to \infty.$$

Two solutions $x = x(t)$ and $y = y(t)$ of (6.3) will be called *geometrically distinct* if $\mathcal{O}(x) \neq \mathcal{O}(y)$, i.e. if $y \neq T_\theta x$ for any θ.

Theorem 6.1 *Suppose $V \in C^1(\mathbf{R}^N, \mathbf{R})$, V satisfies (6.6), (6.7) and $\sigma(A_\infty) \cap \{k^2 : k = 0, 1, 2, \ldots\} = \emptyset$.*
(i) If $i^-(A_0) > i^-(A_\infty)$, $E^+(A_0) + E^-(A_\infty) = \mathbf{R}^N$, $V|_{E^-(A_\infty)} \leq V(0)$ and $V'(x) \neq 0$ whenever $V(x) > V(0)$, then (6.3) has at least $\frac{1}{2}(i^-(A_0) - i^-(A_\infty))$ geometrically distinct nonconstant solutions.
(ii) If $i^-(A_0) + i^0(A_0) < i^-(A_\infty)$, $E^-(A_0) + E^+(A_\infty) = \mathbf{R}^N$, $V|_{E^+(A_\infty)} \geq V(0)$ and $V'(x) \neq 0$ whenever $V(x) < V(0)$, then (6.3) has at least $\frac{1}{2}(i^-(A_\infty) - i^-(A_0) - i^0(A_0))$ geometrically distinct nonconstant solutions.

Proof Assume without loss of generality that $V(0) = 0$.

(i) We will use Theorem 5.11 with $E_n = E$ and $d_n = 0$. So $i_{F,\mathcal{E}} \equiv i_F$. According to Lemma 3.3, Φ satisfies (PS). Set $2l := i_0^-(A_0)$, $2m := i_0^-(A_\infty)$, $F := E^+(A_0)$, $\widetilde{F} := E^-(A_\infty)$, $X := E^+(L_\infty|_{\widetilde{E}})$, $Y := E^-(L_0|_{\widetilde{E}})$ (see (3.12) for the definition of L_0 and L_∞). Observe that since $V|_{E^-(A_\infty)} \leq 0$ and $V|_{E^+(A_0) \cap \partial B_\rho} > 0$ if ρ is small enough (cf. (6.6)), $E^-(A_\infty) \cap E^+(A_0) = \{0\}$. Therefore $F + \widetilde{F} = F \oplus \widetilde{F} = E^G \cong \mathbf{R}^N$. Furthermore, $\dim Y = 2l$ and $\operatorname{codim}_{\widetilde{E}} X = 2m$. Since $E^-(L_0) = E^+(A_0) \oplus E^-(L_0|_{\widetilde{E}}) = F \oplus Y$ and $E^+(L_\infty) = E^-(A_\infty) \oplus E^+(L_\infty|_{\widetilde{E}}) = \widetilde{F} \oplus X$, we see as in the proof of Theorem 3.1 (using Lemma 3.2) that $\Phi|_{(F \oplus Y) \cap \partial B_\rho} \leq \beta$ and $\Phi|_{\widetilde{F} \oplus X} \geq \alpha$ for suitably chosen $\rho > 0$ and $\alpha < \beta < 0$. Also, $V|_{E^-(A_\infty)} \leq 0$ implies $\Phi|_{\widetilde{F}} \geq 0$. If $x \in E^G$, then x is a constant function, and therefore $\nabla\Phi(x) = 0$ if and only if $V'(x) = 0$. So $V(x) \leq 0$ and $\Phi(x) \geq 0$. It follows now from Theorem 5.11 that (6.3) has at least $\frac{1}{2}(i_0^-(A_0) - i_0^-(A_\infty))$ geometrically distinct nonconstant solutions. Finally, since $\dim E^+(A_0) + \dim E^-(A_\infty) = N$ and A_∞ is invertible, it follows that $\dim E^+(A_0) = \dim E^+(A_\infty)$. Hence $i^-(A_0) - i^-(A_\infty) = i_0^-(A_0) - i_0^-(A_\infty)$ according to (6.4).

(ii) Here we will apply Theorem 5.11 to $-\Phi$. Let

$$E_n := \{x \in E : x(t) = a_0 + \sum_{k=1}^{n} a_k \cos kt + b_k \sin kt\},$$

$2d_n := \dim \widetilde{E}_n = 2Nn$ and $\mathcal{E} := \{E_n, d_n\}_{n=1}^\infty$. Define $2\widetilde{l} := i_0^-(A_0) + i_0^0(A_0)$. $2\widetilde{m} := i_0^-(A_\infty)$, $F := E^-(A_0)$, $\widetilde{F} := E^+(A_\infty)$, $X := E^-(L_\infty|_{\widetilde{E}})$, $Y := E^+(L_0|_{\widetilde{E}})$. Now a combination of the argument above with the one in the proof of (ii) of Theorem 3.1 shows that (6.3) has at least $\frac{1}{2}(i_0^-(A_\infty) - i_0^-(A_0) - i_0^0(A_0))$ geometrically distinct nonconstant solutions. Since $\dim E^-(A_0) + \dim E^+(A_\infty) = N$, $\dim E^+(A_0) + \dim E^0(A_0) = \dim E^+(A_\infty)$ and it follows from (6.4), (6.5) that $i^-(A_\infty) - i^-(A_0) - i^0(A_0) = i_0^-(A_\infty) - i_0^-(A_0) - i_0^0(A_0)$. □

Related results may be found in [8].

Let us now turn our attention to the problem

$$\dot{z} = JH'(z), \qquad z(t+2\pi) = z(t). \tag{6.8}$$

We see as in (6.1), (6.2) that there is a one-to-one correspondence between T-periodic solutions for the system $\dot{z} = JH'(z)$ and 2π-periodic solutions for $\dot{z} = \lambda JH'(z)$, where $\lambda := T/2\pi$. Therefore we only consider (6.8). Let $E := H^{1/2}(S^1, \mathbf{R}^{2N})$,

$$\Phi(z) := \frac{1}{2}\int_0^{2\pi} (-J\dot{z}\cdot z)\, dt - \int_0^{2\pi} H(z)\, dt$$

and $(T_\theta z)(t) := z(t+\theta)$. Then $\Phi(T_\theta z) = \Phi(z)$ for all z, θ,

$$E^G = \{z \in E : z(t) = a_0\} \cong \mathbf{R}^{2N}$$

(cf. (4.2)) and

$$E = E^G \oplus \widetilde{E},$$

where as before $\widetilde{E}$ is the subspace of functions having mean value zero.

Let A be a symmetric $2N \times 2N$ matrix and

$$i_0^-(A) := i^-(A) - (M^+(A) - N) = \sum_{k=1}^{\infty}(M^-(T_k(A)) - 2N),$$

$$i_0^+(A) := i^+(A) - (M^-(A) - N) = \sum_{k=1}^{\infty}(M^+(T_k(A)) - 2N)$$

and

$$i_0^0(A) := i^0(A) - M^0(A) = \sum_{k=1}^{\infty} M^0(T_k(A))$$

(cf. (4.7)-(4.9)). Conditions (4.11), (4.12) now read

$$H(z) = \frac{1}{2}A_0 z\cdot z + G_0(z), \text{ where } G_0'(z) = o(|z|) \text{ as } z \to 0 \tag{6.9}$$

and

$$H(z) = \frac{1}{2}A_\infty z\cdot z + G_\infty(z), \text{ where } G_\infty'(z) = o(|z|) \text{ as } |z| \to \infty. \tag{6.10}$$

Theorem 6.2 *Suppose $H \in C^1(\mathbf{R}^{2N}, \mathbf{R})$, H satisfies (6.9), (6.10) and $\sigma(JA_\infty) \cap i\mathbf{Z} = \emptyset$.*

(i) If $i^-(A_0) > i^-(A_\infty)$, $E^+(A_0) + E^-(A_\infty) = \mathbf{R}^{2N}$, $H|_{E^-(A_\infty)} \le H(0)$ and $H'(z) \ne 0$ whenever $H(z) > H(0)$, then (6.8) has at least $\frac{1}{2}(i^-(A_0) - i^-(A_\infty))$ geometrically distinct nonconstant solutions.

(ii) If $i^+(A_0) > i^+(A_\infty)$, $E^-(A_0) + E^+(A_\infty) = \mathbf{R}^{2N}$, $H|_{E^+(A_\infty)} \ge H(0)$ and $H'(z) \ne 0$ whenever $H(z) < H(0)$, then (6.8) has at least $\frac{1}{2}(i^+(A_0) - i^+(A_\infty))$ geometrically distinct nonconstant solutions.

Proof As usual, we may assume $H(0) = 0$. We only prove part (i) because the same argument applied to $-\Phi$ gives (ii).

Let

$$E_n := \{z \in E : z(t) = a_0 + \sum_{k=1}^{n} a_k \cos kt + b_k \sin kt\},$$

$2d_n := \frac{1}{2}\dim \widetilde{E}_n = 2Nn$ and $\mathcal{E} := \{E_n, d_n\}_{n=1}^{\infty}$. Let further $2l := i_0^-(A_0)$, $2m := i_0^-(A_\infty)$, $F := E^+(A_0)$, $\widetilde{F} := E^-(A_\infty)$, $X := E^+(L_\infty|_{\widetilde{E}})$, $Y := E^-(L_0|_{\widetilde{E}})$ (L_0 and L_∞ are defined in (4.14)).

We will verify the hypotheses of Theorem 5.11. As in the proof of Theorem 6.1 we see that $F + \widetilde{F} = F \oplus \widetilde{F} = E^G \cong \mathbf{R}^{2N}$. Moreover, Φ satisfies (PS)* according to Lemma 4.3 and using Lemma 4.2 we see as before that (iii) and (iv) of Theorem 5.11 are satisfied ($\Phi|_{\widetilde{F}} \geq 0$ because $H|_{E^-(A_\infty)} \leq 0$). For n large enough we have

$$\begin{aligned}\operatorname{codim}_{\widetilde{E}_n} X_n &= \operatorname{codim}_{\widetilde{E}_n} E^+(L_\infty|_{\widetilde{E}}) \cap E_n = \dim E^-(L_\infty|_{\widetilde{E}}) \cap E_n \\ &= \sum_{k=1}^{n}(M^-(T_k(A_\infty)) - 2N) + 2Nn = 2(m + d_n)\end{aligned}$$

and similarly,

$$\dim Y_n = \dim E^-(L_0|_{\widetilde{E}}) \cap E_n = 2(l + d_n).$$

Hence (ii) of Theorem 5.11. Since $H(z) \leq 0$ whenever $H'(z) = 0$, $\Phi(z) \geq 0$ for each $z \in E^G$ such that $\nabla\Phi(z) = 0$. This gives (i). Finally, since $\dim E^+(A_0) + \dim E^-(A_\infty) = 2N$ and A_∞ is invertible, $M^+(A_0) = M^+(A_\infty)$ and $i^-(A_0) - i^-(A_\infty) = i_0^-(A_0) - i_0^-(A_\infty)$. □

The above theorem is due to Degiovanni and Olian Fannio [19] and it contains an earlier result by Benci [9, Theorem 5.1] (see also [30]) as a special case (in [9, 30] it was assumed that $H(z) \geq 0$). While we have employed the limit index theory of Section 5, the argument in [19] uses cohomological index theories [20, 21, 22] together with a variant of Benci's pseudoindex [9].

Remark 6.3 According to (6.4), $i_0^-(\lambda^2 A_0) = i_0^-(\lambda^2 A_\infty) = 0$ whenever λ is sufficiently small. So it follows from (6.1), (6.2) that given a potential V, there exists $T_0 > 0$ such that no solution of period $T < T_0$ can be obtained with the aid of Theorem 6.1. This is consistent with the fact that if V' is Lipschitz continuous with Lipschitz constant L, then each nonconstant solution of (6.1) has period $T \geq 2\pi/\sqrt{L}$ [10, Theorem 4.3], [46] (in [10, 46] it is shown that $T \geq 2\pi/L$ for first order systems; however, a simple modification of the argument in [10] gives $T \geq 2\pi/\sqrt{L}$ for systems of second order). A similar remark is easily seen to apply to Hamiltonian systems.

On the other hand, it is easy to give examples of V such that the hypotheses of Theorem 6.1 are satisfied for $\lambda^2 V$ with large λ. So (6.1) will have nonconstant solutions of large period T; however, T need not be the minimal period. Let us in particular consider (6.1) with $N = 1$ (a single equation). Suppose $V \geq 0$ and $0 < A_0 < A_\infty$.

Then $i^-(\lambda^2 A_\infty) > i^-(\lambda^2 A_0) + i^0(\lambda^2 A)$ if λ is large enough, and it follows from (ii) of Theorem 6.1 that (6.1) has a nonconstant periodic solution provided $\lambda\sqrt{A_\infty} \notin \mathbf{N}$. Moreover, $i^-(\lambda^2 A_\infty) - i^-(\lambda^2 A_0) \to \infty$ as $\lambda \to \infty$ and $i^0(\lambda^2 A_0) \leq 2$. So the number of geometrically distinct nonconstant T-periodic solutions tends to infinity with T. However, this does not necessarily mean that (6.1) has solutions of arbitrarily large minimal period. In fact, if $A_0 < V'(x)/x < A_\infty$ for all $x \neq 0$, then all nontrivial solutions of (6.1) have minimal period $T_{\min} \in (2\pi/\sqrt{A_\infty}, 2\pi/\sqrt{A_0})$. Indeed, since $V'(x)/x > 0$ for all $x \neq 0$ and $V(x) \to \infty$ as $x \to \pm\infty$, it follows from the phase-plane analysis that if a solution $\overline{x} \neq 0$ has minimal period T, then $\overline{x}(t_0) = 0$ for some t_0 and $\overline{x}(t)$ has exactly one zero in the interval $(t_0, t_0 + T)$ (see [17, Section I.29] for a detailed argument). Now the Sturm comparison principle applied to the equation $\ddot{x} + A(t)x = 0$, where $A(t) := V'(\overline{x}(t))/\overline{x}(t)$, shows that $T \in (2\pi/\sqrt{A_\infty}, 2\pi/\sqrt{A_0})$.

Remark 6.4 The hypotheses in Theorems 6.1 and 6.2 imply that $E^\pm(A_0)$ and $E^\mp(A_\infty)$ are complementary subspaces of respectively $\mathbf{R}^N$ and $\mathbf{R}^{2N}$. This condition is rather restrictive because in a sense it says that A_0 and A_∞ have similar spectral properties. Perhaps even more restrictive is the assumption that there are no constant solutions on positive (or negative) levels of the functional Φ. It would therefore be interesting to see what can be said about the existence and the number of nonconstant solutions if one removes the conditions on the spectrum of A_0 and A_∞ and on the nonexistence of constant solutions at the levels c_j. As we have already mentioned, we are aware of only one result [7] - for a different problem with a different symmetry group - where it is not assumed that $K_{c_j} \cap E^G = \emptyset$.

References

[1] H. Amann and E. Zehnder, *Nontrivial solutions for a class of nonresonance problems and applications to nonlinear differential equations*, Ann. Scuola Norm. Sup. Pisa (4) 7 (1980), 539-603.

[2] ——, *Periodic solutions of asymptotically linear Hamiltonian systems*, Manuscr. Math. 32 (1980), 149-189.

[3] A. Bahri and H. Berestycki, *Existence of forced oscillations for some nonlinear differential equations*, Comm. Pure Appl. Math. 37 (1984), 403-442.

[4] ——, *Forced vibrations of superquadratic Hamiltonian systems*, Acta Math. 152 (1984), 143-197.

[5] P. Bartolo, V. Benci and D. Fortunato, *Abstract critical point theorems and applications to some nonlinear problems with "strong" resonance at infinity*, Nonl. Anal. TMA 7 (1983), 981-1012.

[6] T. Bartsch, *Topological Methods for Variational Problems with Symmetries*, Lecture Notes in Mathematics 1560, Springer-Verlag, Berlin, 1993.

[7] T. Bartsch and Z.Q. Wang, *Periodic solutions of even Hamiltonian systems on the torus*, Math. Z., to appear.

[8] V. Benci, *A geometrical index for the group S^1 and some applications to the study of periodic solutions of ordinary differential equations*, Comm. Pure Appl. Math. 34 (1981), 393-432.

[9] ——, *On critical point theory of indefinite functionals in the presence of symmetries*, Trans. Amer. Math. Soc. 274 (1982), 533-572.

[10] H. Berestycki, J.M. Lasry, G. Mancini and B. Ruf, *Existence of multiple periodic orbits on star-shaped Hamiltonian surfaces*, Comm. Pure Appl. Math. 38 (1985), 253-289.

[11] M.L. Bertotti, *Forced oscillations of asymptotically linear Hamiltonian systems*, Boll. Un. Mat. Ital. (7) 1-B (1987), 729-740.

[12] K. Borsuk, *Theory of Retracts*, Monografie Matematyczne, Warsaw, 1967.

[13] K.C. Chang, *Infinite Dimensional Morse Theory and Multiple Solution Problem*, Birkhäuser, Boston, 1993.

[14] D.C. Clark, *A variant of the Ljusternik-Schnirelmann theory*, Indiana J. Math. 22 (1973), 65-74.

[15] ——, *Periodic solutions of variational systems of ordinary differential equations*, J. Diff. Eq. 28 (1978), 354-358.

[16] C.V. Coffman, *A minimum-maximum principle for a class of nonlinear integral equations*, J. Analyse Math. 22 (1972), 391-419.

[17] L. Collatz, *Differential Equations*, Wiley, Chichester, 1986.

[18] C.C. Conley and E. Zehnder, *Morse type index theory for flows and periodic solutions for Hamiltonian equations*, Comm. Pure Appl. Math. 37 (1984), 207-253.

[19] M. Degiovanni and L. Olian Fannio, *Multiple periodic solutions of asymptotically linear Hamiltonian systems*, Nonl. Anal. TMA 26 (1996), 1437-1446.

[20] E.R. Fadell and S. Husseini, *Relative cohomological index theories*, Advances in Math. 64 (1987), 1-31.

[21] E.R. Fadell, S.Y. Husseini and P.H. Rabinowitz, *Borsuk-Ulam theorems for arbitrary S^1-actions and applications*, Trans. Amer. Math. Soc. 274 (1982), 345-360.

[22] E.R. Fadell and P.H. Rabinowitz, *Generalized cohomological index theories for Lie group actions with an application to bifurcation questions for Hamiltonian systems*, Inv. Math. 45 (1978), 139-174.

[23] G. Fei, *Relative Morse index and its application to Hamiltonian systems in the presence of symmetries*, J. Diff. Eq. 122 (1995), 302-315.

[24] G. Fournier, D. Lupo, M. Ramos and M. Willem, *Limit relative category and critical point theory.* In: Dynamics Reported 3 (New Series), C.K.R.T. Jones, U. Kirchgraber and H.O. Walther eds., Springer-Verlag, Berlin, 1994, pp. 1-24.

[25] M.A. Krasnoselskii, *Topological Methods in the Theory of Nonlinear Integral Equations*, Pergamon Press, Oxford, 1964.

[26] W. Kryszewski and A. Szulkin, *An infinite dimensional Morse theory with applications*, Trans. Amer. Math. Soc., to appear.

[27] S.J. Li and J.Q. Liu, *Some existence theorems on multiple critical points and their applications*, Kexue Tongbao 17 (1984), 1025-1027 (in Chinese).

[28] ——, *Morse theory and asymptotic linear Hamiltonian system*, J. Diff. Eq. 78 (1989), 53-73.

[29] S.J. Li and A. Szulkin, *Periodic solutions for a class of nonautonomous Hamiltonian systems*, J. Diff. Eq. 112 (1994), 226-238.

[30] Y.Q. Li, *A limit index theory and its applications*, Nonl. Anal. TMA 25 (1995), 1371-1389.

[31] Y. Long, *Maslov-type index, degenerate critical points, and asymptotically linear Hamiltonian systems*, Science in China (Series A) 33 (1990), 1409-1419.

[32] Y. Long and E. Zehnder, *Morse-theory for forced oscillations of asymptotically linear Hamiltonian systems.* In: Stochastic Processes, Physics and Geometry, S. Albeverio, G. Casati, U. Cattaneo, D. Merlini and A. Moresi eds., World Sci. Publ., Singapore, 1990, pp. 528-563.

[33] W. Marzantowicz, *A Borsuk-Ulam theorem for orthogonal T^k and Z_p^r actions and applications*, J. Math. Anal. Appl. 137 (1989), 99-121.

[34] J. Mawhin and M. Willem, *Critical Point Theory and Hamiltonian Systems*, Springer-Verlag, New York, 1989.

[35] L. Nirenberg, *Comments on nonlinear problems*, Le Matematiche 36 (1981), 109-119.

[36] P.H. Rabinowitz, *Some aspects of nonlinear eigenvalue problems*, Rocky Mt. J. Math. 3 (1973), 161-202.

[37] ——, *Some minimax theorems and applications to nonlinear partial differential equations.* In: Nonlinear Analysis, A collection of papers in honor of Erich H. Rothe, L. Cesari, R. Kannan and H.F. Weinberger eds., Academic Press, New York, 1978, pp. 161-177.

[38] ——, *Minimax Methods in Critical Point Theory with Applications to Differential Equations*, CBMS 65, Amer. Math. Soc., Providence, R.I., 1986.

[39] M. Struwe, *Variational Methods*, Springer-Verlag, Berlin, 1990.

[40] A. Szulkin, *Cohomology and Morse theory for strongly indefinite functionals*, Math. Z. 209 (1992), 375-418.

[41] ——, *Bifurcation for strongly indefinite functionals and a Liapunov type theorem for Hamiltonian systems*, Diff. Int. Eq. 7 (1994), 217-234.

[42] K. Thews, *Multiple solutions for elliptic boundary value problems with odd nonlinearities*, Math. Z. 163 (1978), 163-175.

[43] ——, *Non-trivial solutions of elliptic equations at resonance*, Proc. Royal Soc. Edinb. 85A (1980), 119-129.

[44] M. Willem, *Lectures on Critical Point Theory*, Trabalho de Mat. 250, Fundação Univ. Brasília, Brasília, 1983.

[45] ——, *Minimax Theorems*, Birkhäuser, Boston, 1996.

[46] J.A. Yorke, *Periods of periodic solutions and the Lipschitz constant*, Proc. Amer. Math. Soc. 22 (1969), 509-512.

Andrzej Szulkin[1]
Department of Mathematics,
Stockholm University,
Stockholm, Sweden
andrzejs@matematik.su.se

[1]Supported in part by the Swedish Natural Science Research Council

I V SKRYPNIK

Topological characteristics of fully nonlinear parabolic problems

Introduction

This paper is devoted to the introduction of topological characteristics for fully nonlinear parabolic problems. For simplicity we consider the problem with a Dirichlet boundary condition for a second order parabolic equation.

Let Ω be a bounded open set in R^n with sufficiently smooth boundary $\partial\Omega$. We consider the initial boundary value problem

$$\frac{\partial u}{\partial t} - F(x,t,u,D^1u,D^2u) = f(x,t), \quad (x,t) \in Q_T = \Omega \times (0,T) \tag{0.1}$$

$$u(x,t) = g(x,t), \quad (x,t) \in \partial\Omega \times (0,T) \tag{0.2}$$

$$u(x,0) = h(x), \quad x \in \Omega \tag{0.3}$$

in the space $W_p^{(4)}(Q_T) = W_p^{4,2}(Q_T)$ for $p > \frac{n+2}{2}$. We assume some smoothness of the function $F(x,t,\xi)$, the parabolicity condition and some compatibility condition for initial and boundary data.

The problem (0.1)–(0.3) is reduced to the operator equation

$$Au = 0 \tag{0.4}$$

in some subspace X of the space $W_p^{(4)}(Q_T)$ with bounded continuous operator A which satisfies the following condition $(S)_+$: for any sequence $u_n \in X$ weakly convergent to u_0, the inequality

$$\overline{\lim_{n\to\infty}} \langle Au_n, u_n - u_0\rangle \le 0 \tag{0.5}$$

implies that the sequence u_n converges strongly to u_0 (notice that in the monograph [9] we denote such a type of condition as condition α).

These properties of the operator A enable us to study the operator equation (0.4) by topological methods which were developed by the author. Namely, for such operators the author introduced in [9] the notion of degree on bounded set $\overline{D}$ with respect to the point $0 \in X^*$ if $Au \neq 0$ for $u \in \partial D$.

The reduction of the problem (0.1)–(0.3) to the operator equation is based on a priori estimates for linear parabolic operators with nonsmooth coefficients which are proved in this paper and which are analogous to well known a priori estimates in Sobolev spaces for operators with smooth coefficients.

In the paper some applications of the developed topological approach are given. In particular we prove results on local solvability of the problem (0.1)–(0.3) and conditional solvability for an arbitrary T.

In case of elliptic essentially nonlinear boundary value problems analogous topological methods were developed in the monograph [10]. The local existence of initial boundary value problems for fully nonlinear parabolic equations was established by another methods in papers of S.I.Hudjaev [2], S.N.Kruzhkov, A.Castro, M.Lopes [4], N.N. Sopolov [12], P.Acquistapace, B.Terreni [1]. The problem of existence in the large has been studied by N.V.Krylov [5], A.Lunardi [7].

This paper is organized as follows. In §1 we formulate some notions and properties connected with a theory of degree for operators satisfying the condition $(S)_+$. Formulation of main assumptions, results and auxiliary propositions for functions from considered functional spaces can be found in §2, §3. In §4 we prove a priori estimates for linear parabolic operators with coefficients from Sobolev spaces on which the reduction of the problem (0.1)–(0.3) to the operator equation (0.4) is based. In §5 we study the properties of nonlinear operator A and introduce a topological characteristic for the problem (0.1)–(0.3). Some applications of topological method to the study of solvability of parabolic problems are contained in §6.

§1. THE DEGREE OF MAPPING OF CLASS $(S)_+$

This paragraph is devoted to the presentation of the theory of degree of a mapping of class $(S)_+$. Introduction, systematic investigation and various applications of this theory are due to the author (see [9,10]). Notice that in [9] only another terminology was used: we used the notion of *the rotation of mapping of class* α which is the same as *the degree of mapping of class* $(S)_+$.

Throughout this paragraph, X stands for a real separable reflexive Banach space, and X^* for its dual. We denote strong and weak convergence by $\to$ and $\rightharpoonup$, respectively; it will be clear from the context in what space the convergence takes place. For arbitrary elements $u \in X$ and $h \in X^*$, we denote by $\langle h,\ u\rangle$ the action of h on u.

We consider a nonlinear operator A defined on a subset $D(A) \subset X$ with values in X^*.

Definition 1.1. We say that an operator A belongs to *the class* $(S)_+$ if for any sequence $u_n \in D(A)$ for which $u_n \rightharpoonup u_0$, the inequality

$$\overline{\lim_{n\to\infty}} \langle Au_n,\ u_n - u_0\rangle \leq 0 \tag{1.1}$$

implies the strong convergence of the sequence u_n to u_0.

Definition 1.2. An operator A is said to be *demicontinuous* if for any sequence $u_n \in D(A)$ strongly converging to $u_0 \in D(A)$ we have $Au_n \rightharpoonup Au_0$.

Definition 1.3. An operator A is said to be *bounded* if it maps an arbitrary bounded subset of $D(A)$ into a bounded subset of X^*.

In the sequel, D will be a bounded open subset of the space X and ∂D its boundary. We assume that $\overline{D} \subset D(A)$.

Let $\{v_i\}$, $i = 1, 2, ...$ be some complete system of the space X and suppose that for every n the elements $v_1, ..., v_n$ are linearly independent. Denote by F_n the linear hull of $v_1, ..., v_n$.

Define for every $n = 1, 2, ...$ the finite–dimensional mapping A_n corresponding to the operator A in the following way:

$$A_n u = \sum_{i=1}^{n} \langle Au, v_i \rangle v_i \quad \text{for } u \in \overline{D}_n, \;\; D_n = D \cap F_n. \tag{1.2}$$

For the so-defined mapping $A_n : \overline{D}_n \to F_n$, the notion of a degree, if $A_n u \neq 0$ for $u \in \partial D_n$, is well known, see [10]. The degree of A_n with respect to $0 \in F_n$ – denoted by deg $(A_n, \overline{D_n}, 0)$ – is an integer–valued characteristique uniquely determined by the following three properties:

d_1) the normalization condition: if $A_n u = u - u_0$, then deg $(A_n, \overline{D_n}, 0) = 1$ if $u_0 \in D_n$, $\deg(A_n, \overline{D}_n, 0) = 0$ if $u_0 \notin \overline{D}_n$;

d_2) the additivity with respect to the domain: if D'_n, D''_n are disjoint open subsets of D_n and $A_n u \neq 0$ for $u \in \overline{D}_n \backslash (D'_n \cup D''_n)$, then

$$\deg(A_n, \overline{D}_n, 0) = \deg(A_n, \overline{D}'_n, 0) + \deg(A_n, \overline{D}''_n, 0);$$

d_3) the invariance of the degree under homotopy: if $h : \overline{D}_n \times [0,1] \to F_n$ is a continuous mapping such that $h(u,t) \neq 0$ for $u \in \partial D_n$, $t \in [0,1]$ and

$$A_n(u) = h(u,0), \;\; A_n^{(1)}(u) = h(u,1) \;\text{ for } u \in \overline{D}_n,$$

then

$$\deg(A_n, \overline{D}_n, 0) = \deg(A_n^{(1)}, \overline{D}_n, 0).$$

We say that the mapping h performs a homotopy between the mappings $A_n, A_n^{(1)}$.

Theorem 1.1. *Let A be a demicontinuous bounded operator which belongs to the class $(S)_+$ and assume that $\overline{D} \subset D(A)$,*

$$Au \neq 0 \;\; \textit{for} \;\; u \in \partial D. \tag{1.3}$$

Then there exists a number N such that for $n \geq N$ the following assertions hold:

1) the equation $A_n u = 0$ has no solution belonging to ∂D_n;

2) the degree $\deg(A_n, \overline{D}_n, 0)$ is defined and independent of n.

Proof. We prove the first assertion by a contradiction. Assume that there is a sequence $u_k \in \partial D_{n_k}$ such that $n_k \to \infty$ as $k \to \infty$ and $A_n u_k = 0$. We may suppose that u_k converges weakly to some $u_0 \in X$. In addition, $A_n u_k = 0$ implies the weak convergence of Au_k to zero. We show that u_k converges strongly to $u_0 \in \partial D$ and $Au_0 = 0$. This suffices for the proof of 1) by virtue of (1.3).

Fix a subsequence $w_k \in F_{n_k}$ such that $w_k \to u_0$. Then

$$\langle Au_k, u_k - u_0\rangle = \langle Au_k, w_k - u_0\rangle + \langle Au_k, u_k - w_k\rangle.$$

The second term of the right–hand side vanishes by the assumption that $A_{n_k}u_k = 0$ and $u_k - w_k \in F_{n_k}$. The first term tends to zero as $k \to \infty$ due to the boundedness of the operator A and the choice of the sequence w_k. Therefore

$$\lim_{k\to\infty} \langle Au_k, u_k - u_0\rangle = 0.$$

Hence, by virtue of the condition $(S)_+$ we obtain the strong convergence of u_k to u_0 which guarantees that $u_0 \in \partial D$ and $Au_0 = 0$.

Assertion 1) ensures that, for n sufficiently large, the degree $\deg(A_n, \overline{D}_n, 0)$ is well defined. We prove the independence of $\deg(A_n, \overline{D}_n, 0)$ on n by using the auxiliary mapping

$$\widetilde{A}_n u = \sum_{i=1}^{n-1} \langle Au, v_i\rangle v_i + \langle h_n, u\rangle v_n,$$

where h_n is some element of the space X^*, satisfying the conditions

$$\langle h_n, v_i\rangle = 0 \text{ for } i < n, \ \langle h_n, v_n\rangle = 1.$$

By the Leray–Schauder Lemma [10] we infer that

$$\deg(A_{n-1}, \overline{D}_{n-1}, 0) = \deg(\widetilde{A}_n, \overline{D}_n, 0), \tag{1.4}$$

the right–hand side is defined for large n, by virtue of the assertion 1) of Theorem.

In order to prove the equality

$$\deg(\widetilde{A}_n, \overline{D}_n, 0) = \deg(A_n, \overline{D}_n, 0) \tag{1.5}$$

by virtue of the property d_3) of the degree of finite–dimensional mappings, it suffices to establish that

$$tA_n u + (1-t)\widetilde{A}_n u \neq 0 \text{ for } u \in \partial D_n, \ t \in [0,1], \tag{1.6}$$

for n large enough. We prove the assertion (1.6) by a contradiction. Let us assume that there exist sequences $\widetilde{u}_k \in \partial D_{n_k}, t_k \in [0,1]$, such that

$$t_k A_{n_k}\widetilde{u}_k + (1-t_k)\widetilde{A}_{n_k}\widetilde{u}_k = 0, \ n_k \to \infty \text{ as } k \to \infty. \tag{1.7}$$

From (1.7) we derive the equalities

$$\begin{aligned} &\langle A\widetilde{u}_k, v_i\rangle = 0 \text{ for } i = 1, ..., n_k - 1, \\ &t_k\langle A\widetilde{u}_k, v_{n_k}\rangle + (1-t_k)\langle h_n, \widetilde{u}_k\rangle = 0. \end{aligned} \tag{1.8}$$

By the assertion 1) we obtain $0 < t_k < 1$ for sufficiently large k. We may assume that $\widetilde{u}_k$ converges weakly to some $\widetilde{u}_0 \in X$ and we choose a sequence $\widetilde{w}_k$ such that $\widetilde{w}_k \in F_{n_k-1}, \widetilde{w}_k \to \widetilde{u}_0$. Then we obtain from (1.8)

$$\langle A\widetilde{u}_k, \widetilde{u}_k - \widetilde{u}_0\rangle = \langle A\widetilde{u}_k, \widetilde{w}_k - \widetilde{u}_0\rangle - \frac{1-t_k}{t_k}\langle h_{n_k}, \widetilde{u}_k\rangle^2. \tag{1.9}$$

The first term on the right–hand side of (1.9) tends to zero if $k \to \infty$ and, hence, by virtue of the condition $(S)_+$ we deduce the strong convergence of $\widetilde{u}_k$ to $\widetilde{u}_0 \in \partial D$. The first equality in (1.8) yields $A\widetilde{u}_0 = 0$, which contradicts the assumption (1.3). This establishes (1.6) and simultanously (1.5). It follows from (1.4) and (1.5) that $\deg(A_n, \overline{D}_n, 0)$ is independent of n for sufficiently large n; this concludes the proof of the theorem.

By Theorem 1.1 $\lim\limits_{n\to\infty} \deg(A_n, \overline{D}_n, 0)$ exists and we denote it by $D\{v_i\}$.

Theorem 1.2 *Suppose that the conditions of Theorem 1.1. are satisfied. Then the limit*

$$D\{v_i\} = \lim_{n\to\infty} \deg(A_n, \overline{D}_n, 0) \tag{1.10}$$

does not depend on the choice of the sequence $\{v_i\}$.

Proof. We have to show that $D\{v_i'\} = D\{v_i\}$ for any sequence $\{v_i'\}$ with the same properties as $\{v_i\}$. We can assume that for each n the system $v_1, ..., v_n, v_1', ..., v_n'$ is linearly independent. Otherwise, an auxiliary system $\{\widetilde{v}_i\}$ can be constructed such that both the systems $v_1, ..., v_n, \widetilde{v}_1, ..., \widetilde{v}_n$ and $v_1', ..., v_1', \widetilde{v}_1, ..., \widetilde{v}_n$ are linearly independent for each n and the proof reduces to establishing the equality $D\{v_i\} = D\{\widetilde{v}_i\}$.

Let L_{2n} be the linear space spanned by the elements $v_1, ..., v_n, v_1', ..., v_n'$, and define for $n = 1, 2, ...$ the finite–dimensional mappings

$$A_{2n,t}u = \sum_{i=1}^{n} \{\langle Au, v_i\rangle v_i + [t\langle Au, v_i'\rangle + (1-t)\langle f_i^{(n)}, u\rangle]v_i'\},$$

where $u \in L_{2n} \cap \overline{D},\ t \in [0,1]$, and $f_i^{(n)}$ is the element of the space X^* satisfying the conditions

$$\langle f_i^{(n)}, v_j\rangle = 0,\ \langle f_i^{(n)}, v_j'\rangle = \delta_{ij},\ j = 1, ..., n,$$

δ_{ij} being the Kronecker symbol.

Analogously to the proof of (1.6) it is possible to verify that for large n

$$A_{2n,t}u \neq 0 \text{ for } u \in \partial(L_{2n} \cap D),\ t \in [0,1]. \tag{1.11}$$

Then by virtue of the property d_3) of the degree of finite–dimensional mappings we obtain

$$\deg(A_{2n,1}, \overline{L_{2n} \cap D}, 0) = \deg(A_{2n,0}, \overline{L_{2n} \cap D}, 0). \tag{1.12}$$

On the other hand, by the virtue of the Leray–Schauder Lemma [10] and (1.11) we have

$$\deg(A_{2n,0}, \overline{L_{2n} \cap D}, 0) = \deg(A_n, \overline{D}_n, 0) \tag{1.13}$$

with $A_n u$ defined by (1.2). From (1.12), (1.13) follows the equality

$$\deg(A_{2n,1}, \overline{L_{2n} \cap D}, 0) = \deg(A_n, \overline{D}_n, 0). \tag{1.14}$$

Taking into account that in the definition of $A_{2n,1}$ and L_{2n} the elements v_i and v_i' play a symmetric role, we have analogously to (1.14)

$$\deg(A_{2n,1}, \overline{L_{2n} \cap D}, 0) = \deg(A_n', \overline{D}_n', 0) \tag{1.15}$$

for n large enough and A_n', D_n' determined by (2.2) where v_i is replaced by v_i'. The assertion of the Theorem follows from (1.14), (1.15).

The above proved Theorems 1.1 and 1.2 justify the following definition.

Definition 1.4. For an operator A satisfying conditions of Theorem 1.1, its degree on the set $\overline{D}$ with respect to the point $0 \in X^*$ is defined as the number

$$\lim_{n \to \infty} \deg(A_n, \overline{D}_n, 0)$$

where A_n, D_n are determined in accordance with (2.2). This degree is denoted by$\mathrm{Deg}(A, \overline{D}, 0)$.

We shall further formulate some properties of the degree just introduced which are needed in this paper. More details concerning the properties of degree and proofs of formulated results can be found in the books [9], [10].

Definition 1.5. Let $A_0, A_1 : \overline{D} \to X^*$ be demicontinuous bounded operators of class $(S)_+$ and $A_0 u \neq 0$, $A_1 u \neq 0$ for $u \in \partial D$. The operators A_0, A_1 are called homotopic on $\overline{D}$ if there is a parametric family of operators $A_t : \overline{D} \to X^*$, $t \in [0,1]$ satisfying the following conditions:

$h_1)$ $A_t u \neq 0$ for $u \in \partial D$; $t \in [0,1]$;

$h_2)$ the operator $\tilde{A} : [0,1] \times \overline{D} \to X^*$, defined by the equality $\tilde{A}(t,u) = A_t u$, is bounded and demicontinuous ;

$h_3)$ for any sequences $u_n \in \partial D$, $t_n \in [0,1]$ such that $t_n \to t_0$, $u_n \rightharpoonup u_0$ and $\lim\limits_{n \to \infty} \langle A_{t_n} u_n, u_n - u_0 \rangle = 0$, u_n converges strongly to u_0.

Theorem 1.3. *Let $A_0 : \overline{D} \to X^*, A_1 : \overline{D} \to X^*$ be demicontinuous bounded operators of class $(S)_+$. Suppose that $A_0 u \neq 0, A_1 u \neq 0$ for $u \in \partial D$ and that the mappings A_0, A_1 are homotopic on $\overline{D}$. Then*

$$\mathrm{Deg}(A_0, \overline{D}, 0) = \mathrm{Deg}(A_1, \overline{D}, 0). \tag{1.16}$$

The application of the theory of degree of mapping to the solvability of the operator equation

$$Au = 0 \tag{1.17}$$

is based on the following nonzero degree principle.

Theorem 1.4. *Let A be an operator satisfying the hypotheses of Theorem 4.1 and suppose that* $\operatorname{Deg}(A, \overline{D}, 0) \neq 0$. *Then the equation (1.17) has a solution in D.*

We mention a simple consequence of Theorems 1.3 and 1.4.

Theorem 1.5. *Let $A_0 : \overline{D} \to X^*$, $A_1 : \overline{D} \to X^*$ be operators satisfying the hypotheses of Theorem 1.3 and suppose that* $\operatorname{Deg}(A_0, \overline{D}, 0) \neq 0$. *Then the equation $A_1 u = 0$ has at least one solution in D.*

We formulate one frequently used condition ensuring that the degree is nonzero, which follows from the normalization condition d_1) for mapping in finite–dimensional space.

Theorem 1.6. *Let A be an operator satisfying the hypotheses of Theorem 1.1 and suppose that* $0 \in D$, $\langle Au, u \rangle \geq 0$ *for* $u \in \partial D$. *Then* $\operatorname{Deg}(A, \overline{D}, 0) = 1$.

The degree of mapping enables us to prove strong convergence of approximate solutions. Let v_i, F_n be the same as in (1.2). An element $u_n \in F_n$ such that

$$\langle Au_n, v_i \rangle = 0, \quad i = 1, ..., n,$$

is called an approximate solution of the equation (1.17).

Theorem 1.7. *Let A be an operator satisfying the hypotheses of Theorem 1.1. Suppose that* $\operatorname{Deg}(A, \overline{D}, 0) \neq 0$ *and that the equation (1.17) has a unique solution u_0 in the domain D. Then there exists a number N_0 such that approximate solutions u_n exist for $n \geq N_0$ and the sequence u_n converges strongly to u_0.*

§2. Formulation of problems and auxiliary propositions

1. Function spaces. Below, we study boundary value problems for nonlinear parabolic equations in a cylindrical domain $Q_T = \Omega \times (0, T) \subset R^{n+1}$, where $T > 0$ is a finite number and Ω is a bounded connected open set in R^n with smooth boundary $\partial\Omega$. We explain now our notation.

$L_p(Q_T)$ is the Banach spaces consisting of all measurable functions on Q_T with a finite norm

$$\|u\|_{p,Q_T} = \left\{ \int_{Q_T} |u(x,t)|^p \, dx dt \right\}^{1/p}$$

where $p > 1$. For a positive even number l the Sobolev space $W_p^{(l)}(Q_T)$ is defined as the space of all functions $u(x,t) \in L_p(Q_T)$ having generalized derivatives of the form $\left(\frac{\partial}{\partial t}\right)^s D^\alpha u(x,t)$ with any s and α satisfying the inequality $2s + |\alpha| \leq l$. Here $\alpha = (\alpha_1, ..., \alpha_n)$ is a multi-index with nonnegative integer components α_i, $|\alpha| = \alpha_1 + ... + \alpha_n$ is its length and we denote

$$D^\alpha u(x,t) = (\frac{\partial}{\partial x_1})^{\alpha_1} ... (\frac{\partial}{\partial x_n})^{\alpha_n} u(x,t); \quad D^k u(x,t) = \{D^\alpha u(x,t) : |\alpha| = k\}.$$

The norm in $W_p^{(l)}(Q_T)$ is defined by the equality

$$\|u\|_{p,Q_T}^{(l)} = \sum_{j=0}^{l/2} |u|_{p,Q_T}^{(2j)}, \quad |u|_{p,Q_T}^{(2j)} = \sum_{2s+|\alpha|=2j} \|(\frac{\partial}{\partial t})^s D^\alpha u\|_{p,Q_T}. \tag{2.1}$$

For a noninteger positive number l the space $W_p^{(l)}(Q_T)$ consists of functions with finite norm

$$\|u\|_{p,Q_T}^{(l)} = \sum_{2s+|\alpha|<l} \|(\frac{\partial}{\partial t})^s D^\alpha u\|_{p,Q_T} + |u|_{p,Q_T}^{(l)}, \tag{2.2}$$

where

$$|u|_{p,Q_T}^{(l)} = \sum_{2s+|\alpha|=[l]} |(\frac{\partial}{\partial t})^s D^\alpha u|_{p,Q_T}^{(l-[l],0)} + \sum_{l-2<2s+|\alpha|<l} |(\frac{\partial}{\partial t})^s D^\alpha u|_{p,Q_T}^{(0,\frac{l-2s-|\alpha|}{2})}$$

and for $0 < \alpha < 1$

$$|v|_{p,Q_T}^{(\lambda,0)} = \left\{ \int_0^T dt \int_\Omega dx \int_\Omega |v(x,t) - v(y,t)|^p \frac{dy}{|x-y|^{n-1+\lambda p}} \right\}^{1/p},$$

$$|v|_{p,Q_T}^{(0,\lambda)} = \left\{ \int_\Omega dx \int_0^T dt \int_0^T |v(x,t) - v(x,\tau)|^p \frac{d\tau}{|t-\tau|^{1+\lambda p}} \right\}^{1/p}.$$

Here and further we denote by $[l]$ the integer part of l.

In the rest of this section $\partial\Omega$ belongs to class C^{l_0} and $l \leq l_0$. For a noninteger positive number l we denote by $C^{(l)}(\overline{Q}_T)$ the Banach space of functions $u(x,t)$ that are continuous in $\overline{Q}_T$ together with all derivatives of the form $(\frac{\partial}{\partial t})^s D^\alpha u(x,t)$ for $2s + |\alpha| < l$ and have a finite norm

$$\|u\|^{(l)}_{Q_T} = \sum_{2s+|\alpha|\le[l]} |(\frac{\partial}{\partial t})^s D^\alpha u|^{(0)}_{Q_T} + |u|^{(l)}_{Q_T}, \tag{2.3}$$

where

$$|v|^{(0)}_{Q_T} = \max_{(x,t)\in\overline{Q}_T} |v(x,t)|,$$

$$|u|^{(l)}_{Q_T} = \sum_{2s+|\alpha|=[l]} |(\frac{\partial}{\partial t})^s D^\alpha u|^{(l-[l],0)}_{Q_T} + \sum_{l<2s+|\alpha|<l+2} |(\frac{\partial}{\partial t})^s D^\alpha u|^{(0,\frac{l-2s-|\alpha|}{2})}_{Q_T},$$

and for $0<\lambda<1$

$$|v|^{(\lambda,0)}_{Q_T} = \sup\left\{\frac{|v(x,t)-v(x',t)|}{|x-x'|^\lambda} : \quad x,\ x'\in\Omega,\ x\neq x',\ t\in(0,T)\right\},$$

$$|v|^{(0,\lambda)}_{Q_T} = \sup\left\{\frac{|v(x,t)-v(x,t')|}{|t-t'|^\lambda} : \quad x\in\Omega,\ t,\ t'\in(0,T),\ t\neq t'\right\}.$$

Finally, let us define spaces $W^{(l)}_p(S_T)$ and $C^{(l)}(S_T)$ with noninteger number $l\in(0,l_0]$ and $S_T=\partial\Omega\times(0,T)$. At each point of the boundary $\partial\Omega$ there exists a tangent hyperplane. Let $n(\xi)$ be the unit vector of the outer normal to $\partial\Omega$ at the point $\xi\in\partial\Omega$. A Cartesian coordinate system $\{y\}$ with origin at ξ whose y_n - axis is directed along $n(\xi)$ is called a local coordinate system at ξ. It is possible to choose a finite family $\{U_i\}$, $i=1,...,I$, of subsets of $\partial\Omega$ such that $\bigcup_{i=1}^{I} U_i=\partial\Omega$ and for some points $\xi_i\in U_i$ and some positive number d the set U_i is given in a local system at the point ξ_i by the equation

$$y_n = h_i(y'),\ \ y'=(y,...,y_{n-1})\in B'(d)=\{y' : |y'|<d\} \tag{2.4}$$

and $h_i\in C^{l_0}(B'(d))$. For a function $u(x,t)$ on S_T we define $u^{(i)}(y',t)$, $(y',t)\in D'_T=B'(d)\times(0,T)$ by the equality $u^{(i)}(y',t)=u(y,t)$ for $y=(y',h_i(y'))$.

We define $W^{(l)}_p(S_T)$ as a space of all functions $u(x,t)$ on S_T such that for every $i=1,...,I$ functions $u^{(i)}(y',t)$ belong to $W^{(l)}_p(D'_T(d))$. Anologously we define the space $C^{(l)}(S_T)$. Norms in these spaces are given respectively by the formulas

$$\|u\|^{(l)}_{p,S_T} = \{\sum_{i=1}^{I} [\|u^{(i)}(y',t)\|^{(l)}_{p,D'_T(d)}]^p\}^{1/p}, \tag{2.5}$$

$$|u|^{(l)}_{S_T} = \max\{|u^{(i)}(y',t)|^{(l)}_{D'_T(d)} :\ i=1,...,I\}, \text{respectively.} \tag{2.6}$$

It is simple to check that the norms corresponding to various coverings of the surface $\partial\Omega$ by sets U_i are equivalent.

The space $W_p^l(\Omega)$ for noninteger l can be understood as a space of all functions $u(x),\ x \in \Omega$ such that $\tilde{u}(x,t) \in W_p^l(Q_1)$, where $\tilde{u}(x,t) = u(x)$ for $x \in \Omega,\ t \in (0,1)$. The norm in $W_p^{(l)}(\Omega)$ can be defined by the formula

$$\|u\|_{p,\Omega}^{(l)} = \|\tilde{u}\|_{p,Q_1}^{(l)}.$$

The study of the solvability of boundary value problems for parabolic equations is based on properties of the boundary values of functions from spaces $W_p^{(l)}(Q_T)$ which are given in the next lemma.

Lemma 2.1. *Assume that $u(x,t) \in W_p^{(l)}(Q_T)$ with positive even number l and $2s + |\alpha| < l - p/2$. Then the trace of the derivative $(\frac{\partial}{\partial t})^s D^\alpha u(x,t)$ on $\Omega_0 = \{(x,t) : x \in \Omega, t = 0\}$ belongs to the space $W_p^k(\Omega_0)$ for $k = l - 2s - |\alpha| - 2/p$ and the inequality*

$$\|(\frac{\partial}{\partial t})^s D^\alpha u(x,t)\|_{p,\Omega_0}^{(k)} \leq C_1 \|u\|_{p,Q_T}^{(l)} \tag{2.7}$$

holds. If $2s + |\alpha| < l - 1/p$, then the trace of the derivative $(\frac{\partial}{\partial t})^s D^\alpha u(x,t)$ for $(x,t) \in S_T$ belongs to the space $W_p^{(j)}(\Omega)$ for $j = l - 2s - |\alpha| - 1/p$ and the inequality

$$\|(\frac{\partial}{\partial t})^s D^\alpha u(x,t)\|_{p,S_T}^{(j)} \leq C_2 \|u\|_{p,Q_T}^{(l)} \tag{2.8}$$

holds. Constants C_1, C_2 in inequalities (2.7), (2.8) do not depend on $u(x,t)$.

The proof of Lemma 2.1 can be found in [3]. We need also analogous estimates depending on small parameter and we formulate these estimates only for a model domain $D_T(r) = B(r) \times (0,T),\ B(r) = \{x \in R^n : |x| < r\}$.

Lemma 2.2. *Assume that $m = n-1, n$ or $n+1$ and denote $G_T^{(m)} = B'(r) \times \{0\}$ for $m = n-1$, $G_T^{(m)} = D'_T(r) = B'(r) \times (0,T)$, $B'(r) = \{x' = (x_1, ..., x_{n-1}) : |x'| < r\}$ for $m = n$, $G_T = D_T(r)$ for $m = n+1$. Assume that l is either even or noninteger positive number, $p > 1$ and $u(x,t)$ is an arbitrary function from the space $W_p^{(l)}(D_T(r))$.*

1) Let a nonnegative integer s and a multi-index α satisfy the conditions

$$\varepsilon_m = 1 - \frac{2s + |\alpha|}{l} - a_m > 0, \quad \textit{where } a_{n+1} = 0,\ a_n = \frac{1}{lp},\ a_{n-1} = \frac{2}{lp}. \tag{2.9}$$

Then

$$\|(\frac{\partial}{\partial t})^s D^\alpha u(x,t)\|_{p,G_T^{(m)}} \leq C_3 h^{l\varepsilon_m} \left\{ |u|_{p,D_T(r)}^{(l)} + h^{-l} \|u\|_{p,D_T(r)} \right\}. \tag{2.10}$$

2) Assume that the inequality (2.9) is fulfilled and let k be a noninteger positive number which satisfies the condition $k < \varepsilon_m l$. Then the following inequality holds:

$$|(\frac{\partial}{\partial t})^s D^\alpha u(x,t)|^{(k)}_{p,G_T^{(m)}} \leq C_4 h^{l\varepsilon_m - k}\left\{|u|^{(l)}_{p,D_T(r)} + h^{-l}\|u\|_{p,D_T(r)}\right\}. \tag{2.11}$$

Last inequality is valid also in cases $m < n+1$ for $k = \varepsilon_m l$.

3) Let j be a nonnegative number such that $j < l - \frac{n+2}{p}$. Then

$$|u|^{(j)}_{D_T(r)} \leq C_5 h^{l-j-\frac{n+2}{p}}\left\{|u|^{(l)}_{p,D_T(r)} + h^{-l}\|u\|_{p,D_T(r)}\right\}. \tag{2.12}$$

In all inequalities (2.10)-(2.12) h is an arbitrary positive number not greater than $min(r, \sqrt{T})$ and C_3, C_4, C_5 are positive constants independent of the factors next to them.

The assertions of Lemma 2.2 follow from [3].

We will use also an interpolation inequality for functions depending on x and t and given in the next lemma.

Lemma 2.3. *For any function $u(x,t) \in W_p^{(l)}(Q_T) \cap L_\infty(Q_T)$ with positive even l, the inequality*

$$\sum_{|\alpha|=j} \|D^\alpha u\|^{r_j}_{r_j,Q_T} \leq C_6 \left\{\sum_{|\alpha|=l} \|D^\alpha u\|^p_{p,Q_T} \cdot \|u\|^{r_j - p}_{\infty,Q_T} + T\|u\|^{r_j}_{\infty,Q_T}\right\} \tag{2.13}$$

holds with $j = 1, ..., l-1$, $r_j = \frac{lp}{j}$ and a constant C_6 independent of $u(x,t)$ and T. Here $\|\cdot\|_{\infty,Q_T}$ is the norm in $L_\infty(Q_T)$.

The inequality (2.13) follows by using the Nirenberg-Gagliardo interpolation inequality [8] for fixed value of t and then integrating over t.

We will study the solvability of a nonlinear parabolic initial-boundary value problem by reducing it to the problem with zero initial conditions. For this purpose we need subspaces of the spaces $W_p^{(l)}(Q_T)$, $W_p^{(l)}(S_T)$, consisting of functions in a certain sense equal to zero on Ω_0.

For a positive even l we define $W_p^{(l),0}(Q_T)$ as a subspace of functions from the space $W_p^{(l)}(Q_T)$ satisfying the zero initial conditions:

$$\frac{\partial^k u(x,t)}{\partial t^k} = 0 \text{ for } (x,t) \in \Omega_0 \tag{2.14}$$

for $k = 0, ..., l/2 - 1$. The space $W_p^{(l),0}(S_T)$ for nonintegral l and $p(l/2 - [l/2]) > 1$ is defined as the subspace of functions $u(x,t) \in W_p^{(l)}(S_T)$ satisfying conditions (2.14) for $k = 0, ..., [l/2]$ and having finite norm

$$\|u\|^{(l),0}_{p,S_T} = \{\sum_{i=1}^{I} \left(|u^{(i)}(y',t)|^{(l)}_{p,D'_T(d)}\right)^p\}^{1/p}, \tag{2.15}$$

where $u^{(i)}(y',t)$, $D'_T(d)$ have the same sense as in (2.5).

The following property of these spaces is known (see [6, § 4, Chapter 4]): if a function $u(x,t) \in W_p^{(l),0}(Q_T)$ is extended by zero into the domain $Q_{-T} = \Omega \times (-T,0)$ then the resulting function $u^{(0)}(x,t)$ belongs to $W_p^{(l)}(Q_{(-T,T)})$. Here by $Q_{(-T,T)} = \Omega \times (-T,T)$ and $W_p^{(l)}(Q_{(T_1,T_2)})$ we understand the space $W_p^{(l)}(Q_{T_1+T_2})$ after the change of variable $t \to t' + T_1$. Analogous property is valid for functions from the space $W_p^{(l),0}(S_T)$.

For $l > 1/p$ we define a space $\overset{\circ}{W}{}_p^{(l),0}(Q_T)$ as a subspace of functions from $W_p^{(l),0}(Q_T)$ which are equal to zero on S_T.

2. Formulation of main assumptions. Denote by N the number $1+n+\frac{n(n+1)}{2}$. We will write the coordinates of a vector $\xi \in R^N$ in the form $\xi = \{\xi_\alpha : |\alpha| \le 2\}$. Everywhere in the sequel we assume that the domain Ω belongs to the class C^4 and that the number p satisfies the condition

$$p > \frac{n+2}{2}. \tag{2.16}$$

We will investigate the solvability in $W_p^{(4)}(Q_T)$ of the nonlinear initial-boundary value problem

$$\frac{\partial u}{\partial t} - F(x,t,u,D^1u,D^2u) = f(x,t), \quad (x,t) \in Q_T, \tag{2.17}$$

$$u(x,t) = g(x,t), \quad (x,t) \in S_T, \tag{2.18}$$

$$u(x,0) = h(x), \quad x \in \Omega. \tag{2.19}$$

We assume that following conditions are satisfied:

F_1) the function $F(x,t,\xi)$ is defined for $(x,t) \in \overline{Q}_T$, $\xi \in R^N$, and has continuous derivatives with respect to all arguments up to third order, $F(x,t,0) \equiv 0$;

F_2) there exists a continuous non increasing function $\nu : \overline{R}^1_+ \to R^1_+$ such that for every $(x,t) \in \overline{Q}_T$, $\xi \in R^N$, $\eta \in R^n$, we have

$$\sum_{|\alpha|=2} F_\alpha(x,t,\xi)\eta^\alpha \ge \nu(|\xi|)|\eta|^2, \tag{2.20}$$

where $R^1_+ = \{s \in R^1 : s > 0\}$, $F_\alpha(x,t,\xi) = \frac{\partial}{\partial \xi_\alpha} F(x,t,\xi)$, $\eta^\alpha = \eta_1^{\alpha_1} ... \eta_n^{\alpha_n}$.

The functions on the right-hand sides of (2.17)-(2.19) satisfy the following assumptions:

$$f(x,t) \in W_p^{(2)}(Q_T), \;\; g(x,t) \in W_p^{(4-1/p)}(S_T), \;\; h(x) \in W_p^{4-2/p}(\Omega). \tag{2.21}$$

We will assume also that these functions satisfy compatibility conditions for $x \in \partial\Omega,\; t = 0$. These conditions consist of the fact that the function $h(x)$ and the derivative $\frac{\partial u(x,t)}{\partial t}$, which can be determined for $t = 0$ by means of the equation and initial condition, must satisfy for $x \in \partial\Omega,\; t = 0$, to boundary condition and the equality which is followed from the boundary condition (2.18) after taking the derivative at t.

We set

$$u^{(1)}(x) = \frac{\partial u(x,t)}{\partial t}\Big|_{t=0}, \;\; x \in \Omega. \tag{2.22}$$

By (2.17) and (2.19), this function is determined by the equality

$$u^{(1)}(x) = F(x, 0, h(x), D^1 h(x), D^2 h(x)) + f(x,0). \tag{2.23}$$

Definition. We will say that the compatibility condition is fulfilled for the problem (2.17)–(2.19) if for $x \in \partial\Omega$ the equalities

$$h(x) = g(x), \tag{2.24}$$

$$u^{(1)}(x) = \frac{\partial g(x,0)}{\partial t} \tag{2.25}$$

are valid.

In the next section we will reduce the problem (2.17)–(2.19) to the problem with zero initial and boundary data. In this case the fulfilment of the compatibility condition follows from the belonging of the right hand sides of the equation to spaces $W_p^{(2),0}(Q_T)$.

3. Reduction to the problem with zero initial data. We start with two elementary lemmas which follow from the embeddings of Sobolev spaces and which will be useful in this section and later.

Lemma 2.5. *Let $p > \frac{n+2}{2}$ and let $\Phi(x,t,v)$ be a function defined for $(x,t) \in Q_T,\; v = (v_1, ..., v_M) \in R^M$, which has continuous derivatives up to second order with respect to all arguments. Assume that $\Phi(x,t,0) \equiv 0$. Then the operator $\tilde{\Phi}$, defined by*

$$[\tilde{\Phi}(v)](x,t) = \Phi(x,t,v(x,t)), \tag{2.26}$$

is a bounded continuous operator from $[W_p^{(2),0}(Q_T)]^M$ into $W_p^2(Q_T)$, $v(x,t) = (v_1(x,t), ..., v_M(x,t))$.

Proof. From Lemmas 2.2, 2.3 we obtain the estimates

$$\left.\begin{aligned} \|v_i(x,t)\|_{Q_T}^{(\delta)} &\le C_8 \|v_i(x,t)\|_{p,Q_T}^{(2)}, \\ \|D^1 v_i(x,t)\|_{2p,Q_T} &\le C_8 \|v_i(x,t)\|_{p,Q_T}^{(2)}, \end{aligned}\right\} \tag{2.27}$$

with $0 < \delta < 2 - \frac{n+2}{p}$ and some constant C_8. The boundedness of operator $\tilde{\Phi}$ defined by (2.26) follows from the calculation of derivatives $\left(\frac{\partial}{\partial t}\right)^s D^\alpha \Phi(x,t,v(x,t))$, $2s + |\alpha| \le 2$ and from (2.27). If we have some sequences $v_i^{(k)}(x,t)$, $k = 1, 2, ...$, strongly convergent to $v_i^{(0)}(x,t)$ in $W_p^{(2)}(Q_T)$, $i = 1, ..., I$, then these sequences are uniformly convergent and by direct calculation we can establish strong convergence in $W_p^{(2)}(Q_T)$ of the sequence $\Phi(x,t,v^{(k)}(x,t))$ to $\Phi(x,t,v^{(0)}(x,t))$. This guarantees continuity of $\tilde{\Phi}$.

We formulate the problem with zero initial and boundary conditions in the following manner: to find a function $u(x,t) \in \overset{\circ}{W}{}_p^{(4),0}(Q_T)$ such that

$$\frac{\partial u}{\partial t} - F'(x,t,u,D^1u,D^2u) = f'(x,t), \quad (x,t) \in Q_T, \tag{2.28}$$

where

$$f'(x,t) \in W_p^{(2),0}(Q_T) \tag{2.29}$$

and the function $F'(x,t,\xi)$ satisfies the condition F_2 and the following smoothness condition:

F_3) $F'(x,t,0) \equiv 0$ and the derivatives

$$F'_\alpha(x,t,\xi) = \frac{\partial}{\partial \xi_\alpha} F'(x,t,\xi) \ \text{ for } \ 0 \le |\alpha| \le 2 \tag{2.30}$$

exist and are continuous functions of variables $(x,t) \in \overline{Q}_T$, $\xi \in R^N$.

F_4) The operator $\tilde{F}'_\alpha$, defined by the equality

$$[\tilde{F}'_\alpha(u)](x,t) = F'_\alpha(x,t,u(x,t),D^1u(x,t),D^2u(x,t)), \ (x,t) \in Q_T, \tag{2.31}$$

is a bounded continuous operator from $W_p^{(4),0}(Q_T)$ into $W_p^{(2),0}(Q_T)$ with p satisfying (2.16).

Now, we proceed to the reduction of the problem (2.17)–(2.19) to the problem with zero initial and boundary conditions if the assumptions of preceding section are satisfied. We construct at first a function $v(x,t) \in W_p^{(4)}(Q_T)$ such that

$$\frac{\partial^k v(x,t)}{\partial t^k}\Big|_{t=0} = u^{(k)}(x), \ \ k = 0, 1, \tag{2.32}$$

where $u^{(0)}(x) = h(x)$ and $u^{(1)}(x)$ is defined by (2.23).

Check that the function $u^{(1)}(x)$ defined by (2.23) belongs to $W_p^{(2-1/2p)}(\Omega)$. It follows from Lemma 2.2 that $f(x,0)$ belongs to this space. From an analogous embedding result for a function depending only on x it follows that the initial function $h(x)$ belongs to $C^{2+\delta}(\overline{\Omega})$ with $0<\delta<2-\frac{n+2}{p}$. Then it is easy to verify that

$$F(x,0,h(x),D^1h(x),D^2h(x))\in W_p^{2-1/2p}(\Omega),$$

therefore, $u^{(1)}(x)\in W_p^{2-1/2p}(\Omega)$. Moreover it is possible to check that the inequalities

$$\|h(x)\|_{p,\Omega}^{(2-1/2p)}\leq R,\quad \|f(x,t)\|_{p,Q_T}^{(2)}\leq R, \tag{2.33}$$

with some constant R imply the estimate

$$\|u^{(1)}(x)\|_{p,\Omega}^{(2-1/2p)}\leq C_0(R) \tag{2.34}$$

with some continuous function $C_0:\overline{R}_+^1\to R^1+$.

By using well known extension results (see [3]) we can now extend the functions $u^{(0)}(x),u^{(1)}(x)$ onto the entire space R^n in such a way that

$$\|u^{(k)}(x,t)\|_{p,R^n}^{(4-2k-1/p)}\leq C_{10}\|u^{(k)}(x,t)\|_{p,\Omega}^{(4-2k-1/p)},\quad k=0,1. \tag{2.35}$$

By virtue of Theorem 4.4 [6] it is possible to construct a function $v_1(x,t)\in W_p^{(4)}(Q_T)$ satisfying the initial conditions (2.32) and the inequality

$$\|v_1(x,t)\|_{p,Q_T}^{(4)}\leq C_{11}\sum_{k=0}^{1}\|u^{(k)}(x)\|_{p,\Omega}^{(4-2k-1/p)}. \tag{2.36}$$

We set

$$u_1(x,t)=u(x,t)-v_1(x,t), \tag{2.37}$$

which is a solution of the boundary value problem

$$\frac{\partial u_1}{\partial t}-F_1(x,t,u_1,D^1u_1,D^2u_1)=f_1(x,t),\quad (x,t)\in Q_T, \tag{2.38}$$

$$u_1(x,t)=g_1(x,t),\quad (x,t)\in S_T, \tag{2.39}$$

where the functions F_1,f_1,g_1 are defined by the equalities:

$$\begin{aligned}F_1(x,t,\xi)=F(x,t,\xi^{(0)}+v_1(x,t),\xi^{(1)}+D^1v_1(x,t),\xi^{(2)}+D^2v_1(x,t))-\\ -F(x,t,v_1(x,t),D^1v_1(x,t),D^2v_1(x,t)),\end{aligned} \tag{2.40}$$

$$f_1(x,t) = f(x,t) - \frac{\partial v_1(x,t)}{\partial t} - F(x,t,v_1(x,t),D^1 v_1(x,t),D^2 v_1(x,t)),$$

$$g_1(x,t) = g(x,t) - v_1(x,t),$$

and $\xi^{(k)} = \{\xi_\alpha : |\alpha| = k\}$.

Lemma 2.6. *Assume that the conditions (2.16), (2.21), F_1, F_2, and the compatibility condition are satisfied and that $u(x,t) \in W_p^{(4)}(Q_T)$ is a solution of the initial boundary value problem (2.17)–(2.19). Then:*

a) the function $F_1(x,t,\xi)$ defined by (2.40) satisfies the conditions F_2 (with some function $\nu_1(s)$ instead of $\nu(s)$) and the conditions F_3, F_4;

b) the functions $u_1(x,t)$, $f_1(x,t)$, $g_1(x,t)$, defined by (2.37), (2.40) satisfy the conditions

$$u_1(x,t) \in W_p^{(4),0}(Q_T),\ \ f_1(x,t) \in W_p^{(2),0}(Q_T),\ \ g_1(x,t) \in W_p^{(4-1/p),0}(S_T). \quad (2.41)$$

Proof. The assertion of the lemma for the function F_1 follows from the conditions on functions F, the embedding $W_p^{(4)}(Q_T) \subset C^{(2+\delta)}(\overline{Q}_T)$, $0 < \delta < 2 - \frac{n+2}{p}$, and Lemma 2.5. Analogously we obtain the inclusions

$$f_1(x,t) \in W_p^{(2)}(Q_T),\ \ g_1(x,t) \in W_p^{(3-1/p)}(S_T). \quad (2.42)$$

By a construction of functions $u^{(1)}(x)$, $v_1(x,t)$ we have

$$\frac{\partial^k u_1(x,t)}{\partial t^k} = 0 \text{ for } x \in \Omega,\ t = 0,\ k = 0,1, \quad (2.43)$$

and this guarantees the first inclusion in (2.41). The equality

$$f_1(x,t) = 0 \text{ for } x \in \Omega,\ t = 0$$

follows from (2.23), (2.32). We have checked the second inclusion in (2.41).

From the equalities (2.24), (2.25) we obtain

$$g_1(x,t) = \frac{\partial g_1(x,t)}{\partial t} = 0 \ \textit{for}\ x \in \partial\Omega,\ t = 0$$

and this guarantees the third inclusion in (2.41). The proof of Lemma is complete.

Define now a function $v_2(x,t)$ as a solution of initial boundary value problem

$$\frac{\partial v_2(x,t)}{\partial t} - \sum_{i=1}^{n} \frac{\partial^2}{\partial x_i^2} v_2(x,t) = 0,\ \ (x,t) \in Q_T, \quad (2.44)$$

$$v_2(x,t) = g_1(x,t), \quad (x,t) \in S_T, \tag{2.45}$$

$$v_2(x,0) = 0, \quad x \in \Omega. \tag{2.46}$$

It is well known [6] that such a solution exists and satisfies $v_2(x,t) \in W_p^{(4),0}(Q_T)$.

We introduce a new unknown function

$$\overline{u}(x,t) = u_1(x,t) - v_2(x,t), \tag{2.47}$$

which is a solution of the equation

$$\frac{\partial \overline{u}(x,t)}{\partial t} - \overline{F}(x,t,\overline{u},D^1\overline{u},D^2\overline{u}) = \overline{f}(x,t), \quad (x,t) \in Q_T, \tag{2.48}$$

where $\overline{F}, \overline{f}$ are defined by the equalities

$$\begin{aligned}\overline{F}(x,t,\xi) = F_1(x,t,\xi^{(0)} + v_2(x,t), \xi^{(1)} + D^1 v_2(x,t), \xi^{(2)} + D^2 v_2(x,t)) - \\ - F_1(x,t,v_2(x,t), D^1 v_2(x,t), D^2 v_2(x,t)),\end{aligned} \tag{2.49}$$

$$\overline{f}(x,t) = f_1(x,t) - \frac{\partial v_2(x,t)}{\partial t} - F_1(x,t,v_2(x,t),D^1 v_2(x,t),D^2 v_2(x,t)).$$

Lemma 2.7. *Assume that the conditions of Lemma 2.6. are satisfied. Then the function $\overline{F}(x,t,\xi)$ satisfies the conditions $F_2 - F_4$ and*

$$\overline{u}(x,t) \in \overset{\circ}{W}{}_p^{(4),0}(Q_T), \quad \overline{f}(x,t) \in W_p^{(2),0}(Q_T). \tag{2.50}$$

The proof is analogous to that of Lemma 2.6.

Lemma 2.8. *Assume that all hypotheses of Lemma 2.6 are satisfied. Let $\overline{u}(x,t) \in \overset{\circ}{W}{}_p^{(4),0}(Q_T)$ be a solution of the equation (2.48) with zero initial and boundary data and let functions $\overline{F}, \overline{f}$ be defined by (2.49). Then the function $u(x,t) = \overline{u}(x,t) + v_1(x,t) + v_2(x,t)$ is the solution of the initial boundary value problem (2.17)–(2.19).*

The proof immediately follows from the definition of the functions $u_1, u_2, \overline{F}, \overline{f}$, and from $v_i(x,t) \in W_p^{(4)}(Q_T)$, $i = 1,2$.

From Lemmas 2.7, 2.8 it follows that it is enough to study the solvability of the problem with zero initial and boundary data.

Remark 2.1. Evidently we could assume by formulation of the problem (2.17)–(2.19) that F satisfies a weaker smoothness condition than F_1. It is possible to modify for this function the condition F_1 or the conditions F_3, F_4.

§3. Formulation of Main Results

We will construct a nonlinear operator $A : \overset{\circ}{W}{}_p^{(4),0}(Q_T) \to [\overset{\circ}{W}{}_p^{(4),0}(Q_T)]^*$ satisfying the condition $(S)+$ and such that the solvability of the equation (2.48) in $\overset{\circ}{W}{}_p^{(4),0}(Q_T)$ is equivalent to the solvability of the operator equation

$$Au = 0, \quad u(x,t) \in \overset{\circ}{W}{}_p^{(4),0}(Q_T). \tag{3.1}$$

By $[\overset{\circ}{W}{}_p^{(4),0}(Q_T)]^*$ we denote the adjont space to $\overset{\circ}{W}{}_p^{(4),0}(Q_T)$.

We will omit the bar in notions of equation (2.48), so we will consider the problem

$$\frac{\partial u}{\partial t} - F(x,t,u,D^1u,D^2u) = f(x,t), (x,t) \in Q_T, \quad u(x,t) \in \overset{\circ}{W}{}_p^{(4),0}(Q_T), \tag{3.2}$$

where

$$f(x,t) \in W_p^{(2),0}(Q_T) \tag{3.3}$$

and the functions F_1 satisfy the conditions $F_2 - F_4$, p satisfies the condition (2.16).

Let us introduce the operators $L(u)$ by the equality

$$L(u)\varphi = \sum_{|\alpha|\le 2} F_\alpha(x,t,u,D^1u,D^2u)D^\alpha\varphi, \quad (x,t) \in Q_T, \tag{3.4}$$

where $u(x,t)$, $\varphi(x,t) \in \overset{\circ}{W}{}_p^{(4),0}(Q_T)$, and $F_\alpha(x,t,\xi)$ are of the same meaning as in (2.20). Clearly, for a fixed function $u(x,t)$ defined by (3.4), $L(u)$ is a linear oprerator with respect to $\varphi(x,t)$.

Lemma 3.1. *Assume that the conditions (2.16), $F_2 - F_4$, are fulfilled. Then there exists a continuous nondecreasing function $\omega_1 : \overline{R}{}_+^1 \to R_+^1$ such that the inequality*

$$\|L(u)\varphi\|_{p,Q_T}^{(2)} \le C\omega_1(\|u\|_{p,Q_T}^{(4)}) \cdot \|\varphi\|_{p,Q_T}^{(4)} \tag{3.5}$$

is valid for arbitrary functions $u(x,t)$, $\varphi(x,t) \in \overset{\circ}{W}{}_p^{(4),0}(Q_T)$.

Proof. Denote $\|u\|_{p,Q_T}^{(4)} = R$. By F_4, the norm in $W_p^{(2)}(Q_T)$ of the coefficients of the linear operator $L(u)$ can be estimated from above by $C'(R)$ with some nondecreasing function $C'(R)$. By using of Lemma 2.5 we obtain the estimate

$$\|L(u)\varphi\|_{p,Q_T}^{(2)} \le C''(R) \quad \text{if} \quad \|\varphi\|_{p,Q_T}^{(4)} = 1$$

with some nondecreasing function $C''(R)$. From the last estimate we immediately obtain the inequality (3.5).

Noting that

$$F(x,t,u,D^1u,D^2u) = \int_0^1 L(su)u\,ds \tag{3.6}$$

we obtain from Lemma 3.1 the next corollary.

Corollary 3.1. *For an arbitrary function* $u(x,t) \in \overset{\circ}{W}{}_p^{(4),0}(Q_T)$ *we have*

$$\|F(x,t,u,D^1u,D^2u)\|^{(2)}_{p,Q_T} \le \omega_1(\|u\|^{(4)}_{p,Q_T})\|u\|^{(4)}_{p,Q_T} \tag{3.7}$$

with the same function ω_1 *as in Lemma 3.1.*

Let us introduce the operator

$$E : \overset{\circ}{W}{}_p^{(4),0}(Q_T) \times \overset{\circ}{W}{}_p^{(2),0}(Q_T) \times [\overset{\circ}{W}{}_p^{(4),0}(Q_T)]^*$$

for $u(x,t),\ \varphi(x,t) \in \overset{\circ}{W}{}_p^{(4),0}(Q_T),\ \ v(x,t) \in \overset{\circ}{W}{}_p^{(4),0}(Q_T)$:

$$\langle E(u,v),\varphi\rangle = \int_{Q_T} \sum_{2s+|\alpha|\le 2} \psi_p\{(\frac{\partial}{\partial t})^s D^\alpha v(x,t)\}(\frac{\partial}{\partial t})^s D^\alpha[\frac{\partial\varphi}{\partial t} - L(u)\varphi]dxdt \tag{3.8}$$

For $\varphi(x,t) \in \overset{\circ}{W}{}_p^{(4),0}(Q_T)$ and a functional $l \in [\overset{\circ}{W}{}_p^{(4),0}(Q_T)]^*$ we will denote by $< l, \varphi >$ the value of l on the function $\varphi(x,t)$. By ψ_p we denote the function $\psi_p,\ R^1 \to R^1$, defined by the formula

$$\psi_p(s) = |s|^{p-2}\cdot s. \tag{3.9}$$

From Lemma 3.2 below follows that the operator E is well defined by the equality (3.8).

Lemma 3.2. *Assume that the hypotheses of Lemma 3.1 are fulfilled. Then there exists a continuous not decreasing function* $\omega_2 : \overline{R}^1_+ \to R^1_+$ *such that inequality*

$$| < E(u,v),\varphi > | \le \omega_2(\|u\|^{(4)}_{p,Q_T})\cdot\{\|v\|^{(2)}_{p,Q_T}\}^{p-1}\cdot\|\varphi\|^{(4)}_{p,Q_T} \tag{3.10}$$

is valid.

Proof. The estimate (3.10) follows immediatelly from Lemma 3.1 and the Hölder inequality.

Let us define the operator

$$A : \overset{\circ}{W}{}_p^{(4),0}(Q_T) \to [\overset{\circ}{W}{}_p^{(4),0}(Q_T)]^*$$

by

$$Au = E(u, \frac{\partial u}{\partial t} - F(x,t,u,D^1u,D^2u) - f) \tag{3.11}$$

Taking into account Corollary 3.1 and Lemma 3.2 we obtain that the operator A is well defined.

Fundamental results are given in next theorems.

Theorem 3.1. *Assume that the conditions (2.16), (3.3), $F_2 - F_4$ are satisfied. Then the operator A defined by (3.11) is bounded continuous and satisfies the condition $(S)_+$ on $\overset{\circ}{W}{}_p^{(4)}(Q_T)$.*

Theorem 3.2. *Assume that the conditions of Theorem 3.1 are fulfilled. Then a function $u(x,t) \in \overset{\circ}{W}{}_p^{(4),0}(Q_T)$ is the solution of the problem (3.2) if and only if it is the solution of the operator equation (3.1) with the operator A defined by (3.11).*

Theorems 3.1, 3.2 will be proved in § 5 on base of a priori estimates and solvability results for linear parabolic problems with coefficients from Sobolev spaces, which will be established in § 4. Also by using the results of § 4 we will prove the following uniqueness theorem.

Theorem 3.3. *Assume that the assumptions of Theorem 3.1 are fulfilled. Then the problem (3.2) can have no more than one solution.*

It is evident that the uniqueness theorem is valid also for the initial-boundary value problem (2.17)-(2.19) if the conditions (2.16), (2.21), F_1, F_2 and the compatibility condition are fulfilled. This follows immediately from the reduction of problem (2.17)-(2.19) given in Section 3 of § 2 to the problem with zero initial and boundary data.

From Theorem 3.1 it follows that we can introduce the degree of a mapping A on base of the result of § 1. Let D be a bounded open set in $\overset{\circ}{W}{}_p^{(4),0}(Q_T)$ such that

$$Au \neq 0 \text{ for } u \in \partial D, \tag{3.12}$$

where ∂D is a boundary of D and the operator A is defined by (3.11). Then from § 1 it follows that the degree of A is defined on the set $\overline{D}$ with respect to the point $0 \in [\overset{\circ}{W}{}_p^{(4),0}(Q_T)]^*$. This degree is an integer number which is denoted by $\mathrm{Deg}(A, \overline{D}, 0)$.

The degree of a mapping A defined by (3.11) gives a topological characteristic for nonlinear parabolic problem (3.2) (or analogously for the problem (2.17)–(2.19)). By using the degree theory it is possible to study the solvability of the problem (3.2) or (2.17)–(2.19) by topological methods.

We formulate below two results which will be proved by topolocigal methods. We formulate local results in connection with the problem (2.17)–(2.19).

Theorem 3.4. *Assume that the conditions (2.16), (2.21), F_1, F_2 and the compability condition are satisfied and let K be a positive constant. Then there exists a positive number T_0 such that the initial boundary value problem (2.17)–(2.19) has a unique solution $u(x,t) \in \overset{\circ}{W}{}_p^{(4),0}(Q_T)$ if the inequalities*

$$\|f(x,t)\|_{p,Q_T}^{(2)} \leq K, \quad \|g(x,t)\|_{p,S_T}^{(4-\frac{1}{p})} \leq K, \quad \|h(x)\|_{p,\Omega}^{(4-\frac{2}{p})} \leq K$$

are satisfied and $0 < t \leq T_0$.

Theorem 3.4 gives a local existence result. It is possible to prove also a global conditional result which is given in next theorem. We include the problem (3.2) in a parametric family of analogous problems with functions $F_\lambda(x,t,\xi)$, $\lambda \in [0,1]$ such that

$$F(x,t,\xi) = F_1(x,t,\xi). \tag{3.13}$$

In particular, this family can be chosen in the following manner:

$$F_\lambda(x,t,\xi) = \lambda F(x,t,\xi) + (1-\lambda) \sum_{|\alpha|\le 2} F_\alpha(x,t,0)\xi_\alpha. \tag{3.14}$$

Theorem 3.5. *Let $F(\lambda,x,t,\xi)$ be a continuous function for $\lambda \in [0,1]$, $x \in \overline{\Omega}$, $t \in [0,T]$, $\xi \in R^N$. Suppose that for every $\lambda \in [0,1]$ the function*

$$F_\lambda(x,t,\xi) = F(\lambda,x,t,\xi). \tag{3.15}$$

satisfies the conditions $F_2 - F_4$ of § 2 and suppose that the conditions (2.16), (3.3), (3.13) are fulfilled. In addition we assume that for a given function $f(x,t)$ there exists a constant R such that for an arbitrary admissible solution $u_\lambda(x,t) \in \overset{\circ}{W}{}_p^{(4),0}(Q_T)$ of the equation

$$\frac{\partial u}{\partial t} - F_\lambda(x,t,u,D^1u,D^2u) = \lambda f(x,t), \quad (x,t) \in Q_T, \tag{3.16}$$

an a priori estimate

$$\|u_\lambda(x,t)\|_{p,Q_T}^{(4)} \le R \tag{3.17}$$

is valid. Then the problem (3.2) has a unique solution $u(x,t) \in \overset{\circ}{W}{}_p^{(4),0}(Q_T)$.

Theorems 3.3, 3.4 and 3.5 will be proved in § 6.

§4. A priori estimates for linear parabolic problems with coefficients from Sobolev spaces

The study of the properties of the operator A introduced in the preceding paragraph is based on the solvability results of this paragraph and on estimates of solutions for a linear parabolic problem with nonsmooth coefficients.

Let us consider the initial boundary value problem

$$Lu \equiv \frac{\partial u}{\partial t} - \sum_{|\alpha|\le 2} a_\alpha(x,t)D^\alpha u = f(x,t), \quad (x,t) \in Q_T, \tag{4.1}$$

$$u(x,t) \in \overset{\circ}{W}{}_p^{(4),\circ}(Q_T). \tag{4.2}$$

For simplicity we consider here the problem with zero initial and boundary data.

We assume that $f(x,t) \in W_p^{(2),0}(Q_T)$ and that the following conditions are satisfied:

a_1) the functions $a_\alpha(x,t)$, $|\alpha| \le 2$, are defined for $(x,t) \in Q_T$ and belong to the space $W_p^{(2),0}(Q_T)$ with p satisfying (2.16);

a_2) there exists a positive constant ν_0 such that for every $(x,t) \in Q_T$, $\eta \in R^n$, the inequality

$$\sum_{|\alpha|=2} a_\alpha(x,t)\eta^\alpha \ge \nu_0|\eta|^2 \tag{4.3}$$

holds.

Theorem 4.1. *Assume that the conditions a_1), a_2), (2.16) are satisfied, $T \in (0,T_0]$, and a bounded domain Ω belongs to the class C^4. Then for an arbitrary function $f(x,t) \in W_p^{(2),0}(Q_T)$ the problem (4.1), (4.2) has a unique solution $u(x,t)$ and for this solution we have*

$$K_1\|f(x,t)\|_{p,Q_T}^{(2)} \le \|u(x,t)\|_{p,Q_T}^{(4)} \le K_2\|f(x,t)\|_{p,Q_T}^{(2)} \tag{4.4}$$

with positive constants K_1, K_2 independent of T and $u(x,t)$.

Remark 4.1. The constants K_1, K_2 in Theorem 4.1 depend only on Ω, T_0, and the coefficients $a_\alpha(x,t), p$. We will denote in this paragraph constants dependent only on these parameters by K_i, $i = 3,4,\dots$.

Proof. The first inequality in (4.4) follows analogously to the proof of Lemma 3.1. The proof of the solvability of the problem (4.1), (4.2) and of the second inequality in (4.4) is analogous to that of Theorem 9.1, Chapter 4 in [6] for the case of smooth coefficients. It is necessary to change some estimates taking into account new conditions on coefficients. So we will omit many details in the proof.

At first Theorem 4.1. will be proved for sufficiently small T_0. In order to prove results for this case we construct a bounded operator R acting from $W_p^{(2),0}(Q_T)$ to $\overset{\circ}{W}{}_p^{(4),0}(Q_T)$ such that for every $\varphi(x,t) \in W_p^{(2),0}(Q_T)$ and $v(x,t) \in \overset{\circ}{W}{}_p^{(4),0}(Q_T)$ we have equalities

$$LR\varphi = \varphi + B_1\varphi, \quad RLv = v + B_2 v \tag{4.5}$$

where B_1, B_2 are bounded operators on the spaces $W_p^{(2),0}(Q_T)$ and $\overset{\circ}{W}{}_p^{(4),0}(Q_T)$, respectively, the norms of which are less then $1/2$ if T_0 is sufficiently small. The operator R is defined analogously to the construction of the operator R' in §9, Chapter IV [6] (only for another functional spaces and zero boundary data);

$$R\varphi = \sum_k \eta^{(k)}(x)v^{(k)}(x,t), \tag{4.6}$$

where $\sum\limits_k$ is a finite sum. For a given positive λ, $\eta^{(k)}(x)$ are C^∞ functions defined on $\overline{\Omega}$ diameters of whose supports are less then λ and such the inequalities

$$\eta^{(k)}(x) \ge 0, \quad |D^\alpha\eta^{(k)}(x)| \le C_{12}\lambda^{-|\alpha|}, \quad |\alpha| \le 4, \tag{4.7}$$

are valid with a fixed constant C_{12}. The functions $v^{(k)}(x,t)$ are defined as solutions of model problems: Cauchy problem if supp $\eta^{(k)}(x)\cap\partial\Omega = 0$ or model initial boundary value problem in a half space if supp $\eta^{(k)}\cap\partial\Omega\neq 0$. By this definition, $v^{(k)}(x,t)\in W_p^{(4),0}(Q_T^{(k)})$, and the inequality

$$|v^{(k)}(x,t)|^{(4)}_{p,Q^{(k)}}\leq C_{13}|\eta^{(k)}(x)\varphi(x,t)|^{(2)}_{p,Q_T^{(k)}} \tag{4.8}$$

is valid, where $Q_T^{(k)}=\Omega^{(k)}\times(0,T)$, $\Omega^{(k)}$ is smooth domain such that supp $\eta^{(k)}(x)\subset\Omega^{(k)}$, and the seminorms in (4.8) is defined by (2.1).

We consider only the evaluation of principal summands in the representation of $B_1\varphi$;

$$B_1'\varphi=\sum_k\sum_{|\alpha|=2}\eta^{(k)}(x)[a_\alpha(x,t)-a_\alpha(\xi^{(k)},0)]D^\alpha v^{(k)}(x,t), \tag{4.9}$$

$$B_1''\varphi=\sum_k\sum_{\substack{\alpha'+\alpha''=\alpha\\|\alpha'|\geq 0}}D^{\alpha'}\eta^{(k)}(x)a_\alpha(x,t)D^{\alpha''}v^{(k)}(x,t),$$

where $\xi^{(k)}$ is some point from the support of the function $\eta^{(k)}(x)$.

We introduce in the space $W_p^{(2m),0}(Q_T)$ a new norm

$$\|u\|^{[2m]}_{p,Q_T}=\{\sum_k\sum_{2s+|\alpha|=2m}\int_{Q_T^{(k)}}|(\frac{\partial}{\partial t})^s D^\alpha u(x,t)|^p dxdt\}^{1/p}, \tag{4.10}$$

which is equavalent to the norm defined by (2.1). We will assume that

$$T=\lambda^2\kappa,\ \ 0<\kappa\leq 1 \tag{4.11}$$

and we will choose λ and κ sufficiently small. Remark also that the choice of functions $\eta^{(k)}(x)$ can be such that

$$\sum_k\chi(\Omega^{(k)})\leq K_0 \tag{4.12}$$

with a constant K_0 independent of λ. In (4.12), $\chi(\Omega^{(k)})$ is the characteristic function of $\Omega^{(k)}$.

In terms of the norm defined by (4.10) we have

$$\|B_1'\varphi\|^{[2]}_{p,Q_T} \le K_3\{\sum_k \sum_{|\alpha|=2} \int_{Q_T^{(k)}} [\eta^{(k)}(x)|a_\alpha(x,t) - a_\alpha(\xi_k,0)|\cdot$$

$$\cdot |\frac{\partial}{\partial t} D^\alpha v^{(k)}(x,t)| + \eta^{(k)}(x)|\frac{\partial a_\alpha(x,t)}{\partial t}| \, |D^\alpha v^{(k)}(x,t)| + \tag{4.13}$$

$$+ \sum_{|\beta'+\beta''+\beta'''|=2} |D^{\beta'}[a_\alpha(x,t) - a_\alpha(\xi_k,0)]|\cdot$$

$$\cdot |D^{\beta''}\eta^{(k)}(x)| \, |D^{\beta'''+\alpha} v^{(k)}(x,t)|]^p \, dxdt\}^{1/p},$$

where $\beta', \beta'', \beta'''$ are arbitrary multi–indices such that $|\beta'| + |\beta''| + |\beta'''| = 2$.

To evaluate the right–hand side of the last inequality we will use the estimate

$$|(\frac{\partial}{\partial t})^s D^\gamma u(x,t)|^{(\delta)}_{Q_T^{(k)}} \le K_5 T^{\frac{1}{2}(4-2s-|\gamma|-\frac{n+2}{p}-\delta)} |u|^{(4)}_{p,Q_T^{(k)}} + \\ + K_5 T^{-\frac{1}{2}(2s+|\gamma|+\frac{n+2}{p}+\delta)} \|u\|^{(4)}_{p,Q_T^{(k)}} \tag{4.14}$$

for $4 - 2s - |\gamma| - \frac{n+2}{p} > 0,\ 0 < \delta < 4 - 2s - |\gamma| - \frac{n+2}{p}$. This estimate follows from Lemma 2.2.

Since $u(x,t) \in W_p^{(4),0}(Q_T^{(k)})$, we obtain the inequalities

$$|u|^{(j)}_{p,Q_T^{(k)}} \le K_6 T^{2-j/2} |u|^{(4)}_{p,Q_T^{(k)}} \ \text{ for } j < 4, \tag{4.15}$$

$$|(\frac{\partial}{\partial t})^s D^\gamma u|^{(0)}_{Q_T^{(k)}} \le K_6 T^{\delta/2} |(\frac{\partial}{\partial t})^s D^\gamma u|^{(\delta)}_{Q_T^{(k)}} \tag{4.16}$$

for s, γ, δ satisfying the same conditions as in the estimate (4.14). The inequalities (4.15), (4.16) follow from Lemma 4.5, Chapter 4 [6].

By using the Lemma 2.3 and the inequalities (4.14)–(4.16) we have the estimate

$$\|(\frac{\partial}{\partial t})^s D^\gamma u(x,t)\|^{2p}_{2p,Q_T^{(k)}} \le K_7 T^{\frac{p}{2}(2-\frac{n+2}{p})} [|u|^{(4)}_{p,Q_T^{(k)}}]^{2p} \tag{4.17}$$

for $s + |\gamma| = 3,\ u(x,t) \in W_p^{(4),0}(Q_T^{(k)})$. We will use the estimates (4.14)–(4.17) for functions $v^{(k)}(x,t)$ and analogous estimates for $\varphi(x,t) \in W_p^{(2),0}(Q_T)$.

Now we are able to evaluate summands on the right hand side of (4.13). From the conditions a_1), (2.16) and Lemma 2.2 it follows that $a_\alpha(x,t) \in C^{(\delta_0)}(\overline{Q}_T)$ with $0 < \delta_0 < 2 - \frac{n+2}{p}$. Using this and the estimates (4.8), (4.15) we evaluate the terms on the right hand side of (4.13) that correspond to the first summand in the integrand and to the terms of the third summand with $\beta' = 0$ by

$$K_8 \lambda^{\delta_0} \|\varphi\|^{[2]}_{p,Q_T}.$$

The terms on the right hand side of (4.13) corresponding to the second summand in the integrand and to the terms of the third summand with $|\beta'| = 2$ are evaluated by using the estimates (4.8), (4.14)–(4.16), by

$$K_9 T^{\frac{1}{2}(2-\frac{n+2}{p})} \|\varphi\|^{[2]}_{p,Q_T}.$$

The terms on the right hand side of (4.13) corresponding to the terms of the third summand with $|\beta'| = 1$ are evaluated by using Lemma 2.3 (for $D^{\beta'} a_\alpha(x,t)$) and the estimates (4.8), (4.14)–(4.17) by

$$K_{10} T^{\frac{1}{4}(2-\frac{n+2}{p})} \|\varphi\|^{[2]}_{p,Q_T}.$$

From the inequality (4.13) and the estimates for the terms on the right hand side of (4.13) we obtain

$$\|B_1'\varphi\|^{[2]}_{p,Q_T} \le K_{11}\{\lambda^{\delta_0} + T^{\frac{1}{4}(2-\frac{n+2}{p})}\}\|\varphi\|^{[2]}_{p,Q_T}, \tag{4.18}$$

and consequently the norm of the operator B_1' in $W^{(2)}_{p,Q_T}$ is less then 1/2 for sufficiently small λ.

Analogously to (4.13) we have the estimate for $B_1''\varphi$;

$$\begin{aligned} \|B_1''\varphi\|^{[2]}_{p,Q_T} \le K_{12}\{\sum_k \sum_{\substack{|\alpha'|+|\alpha''|=2 \\ |\alpha'|\ge 1}} \int_{Q_T^{(k)}} [|D^{\alpha'}\eta^{(k)}(x)|\, |a_\alpha(x,t)| \cdot \\ \cdot |\frac{\partial}{\partial t} D^{\alpha''} v^{(k)}(x,t)| + |D^{\alpha'}\eta^{(k)}(x)|\, |\frac{\partial a_\alpha(x,t)}{\partial t}|\, |D^{\alpha''} v^{(k)}(x,t)| + \\ + \sum_{|\beta'+\beta''+\beta'''|=2} |D^{\beta'} a_\alpha(x,t)|\, |D^{\beta''+\alpha'}\eta^{(k)}(x)|\, |D^{\beta'''+\alpha''} v^{(k)}(x,t)|]^p dxdt\}^{1/p} \end{aligned} \tag{4.19}$$

We evaluate the terms on the right hand side of (4.19) corresponding to the first summand in the integrand and to the terms of the third summand with $\beta' = 0$ by using the boundedness of $a_\alpha(x,t)$ and the inequality (4.15) and we obtain that these terms are less then

$$K_{13}(\frac{\sqrt{T}}{\lambda} + \frac{T}{\lambda^2})\|\varphi\|^{[2]}_{p,Q_T}.$$

Evaluating other terms on the right hand side of (4.19) analogously to the evaluation of corresponding terms from (4.13) and using (4.11) we obtain

$$\|B_1''\varphi\|^{[2]}_{p,Q_T} \le K_{14}\{\sqrt{\kappa} + T^{\frac{1}{4}(2-\frac{n+2}{p})}\}\|\varphi\|^{[2]}_{p,Q_T}. \tag{4.20}$$

From last inequality it follows that the norm of the operator B_1'' is less then 1/2 for sufficiently small κ, T_0.

We thus prove that the norms of operators B_1, B_2 from (4.5) are less then $1/2$ for small κ, T_0. By using a discussion analogous to that of the proof of Theorem 9.1 in [6] we can prove the existence and the estimate (4.4) for the problem (4.1), (4.2) at first for small T_0 and then for an arbitrary T_0. The proof of Theorem 4.1 is complete.

§5. Proofs of Theorems 3.1, 3.2

To study the properties of the operator A defined by (3.11) we will use the following auxiliary proposition in which we keep our assumptions concerning Ω.

Lemma 5.1. *Assume that $G(x,t,v)$ is a continuous function of the variables $(x,t) \in \overline{Q_T}$, $v = (v_1, ..., v_N) \in R^N$, such that the operator $\widetilde{G}$, defined by the equality*

$$\widetilde{G}\left(v_1(x,t), ..., v_N(x,t)\right) = G(x,t,v_1(x,t), ..., v_N(x,t)), \tag{5.1}$$

is bounded from $\left[W_p^{(2)}(Q_T)\right]^N$ into $W_p^{(2),0}(Q_T)$ with p satisfying (2.16). Then for arbitrary sequences $v_i^{(n)}(x,t)$, $i = 0, 1, ..., N$, $n = 1, 2, ...$, weakly convergent in $W_p^{(2)}(Q_T)$ to $v_i^{(0)}(x,t)$, respectively, the sequence

$$\begin{aligned} r_n(x,t) = \left[\widetilde{G}\left(v_1^{(n)}(x,t), ..., v_N^{(n)}(x,t)\right) - \widetilde{G}\left(v_1^{(0)}(x,t), ..., v_N^{(0)}(x,t)\right)\right] \cdot \\ \cdot \left[v_0^{(n)}(x,t) - v_0^{(0)}(x,t)\right] \end{aligned} \tag{5.2}$$

converges strongly to zero in $W_p^{(2)}(Q_T)$.

Proof. Taking into account the compactness of the embedding of $W_p^1(Q_T)$ into $L_p(Q_T)$ and the estimate (2.12) it is possible to obtain strong convergence of $v_i^{(n)}(x,t)$ in $C^{(\delta)}(\overline{Q_T})$ for $0 < \delta < 2 - \frac{n+2}{p}$. Using the continuity of the function $G(x,t,v)$ we obtain immediately strong convergence of $r_n(x,t)$ to zero in $L_p(Q_T)$ and uniform convergence to zero of

$$\overline{G_n}(x,t) = \widetilde{G}\left(v_1^{(n)}(x,t), ..., v_N^{(n)}(x,t)\right) - \widetilde{G}\left(v_1^{(0)}(x,t), ..., v_N^{(0)}(x,t)\right).$$

In the equality

$$\frac{\partial r_n(x,t)}{\partial t} = \frac{\partial \overline{G}_n(x,t)}{\partial t} \cdot \left[v_0^{(n)}(x,t) - v_0^{(0)}(x,t)\right] + \left[\frac{\partial v_0^{(n)}(x,t)}{\partial t} - \frac{\partial v_0^{(0)}(x,t)}{\partial t}\right] \overline{G}_n(x,t) \tag{5.3}$$

both summands on the right hand side tend to zero in $L_p(Q_T)$ since the first factors of these summands are bounded in $L_p(Q_T)$ and the second factors uniformly converge to zero. Analogously we verify the strong convergence in $L_p(Q_T)$ of the derivatives $\frac{\partial r_n(x,t)}{\partial x_j}$, $j = 1, ..., n$.

For the second order derivatives with respect to x we have

$$D^\alpha r_n(x,t) = D^\alpha \overline{G_n}(x,t) \cdot \left[v_0^{(n)}(x,t) - v_0^{(0)}(x,t)\right] + \left[D^\alpha v_0^{(n)}(x,t) - D^\alpha v_0^{(0)}(x,t)\right] \cdot$$

$$\cdot \overline{G_n}(x,t) + \sum_{\substack{|\alpha'|=|\alpha''|=1 \\ \alpha'+\alpha''=\alpha}} D^{\alpha'} \overline{G_n}(x,t) \cdot \left[D^{\alpha''} v_0^{(n)}(x,t) - D^{\alpha''} v_0^{(0)}(x,t)\right]. \tag{5.4}$$

The strong convergence in $L_p(Q_T)$ to zero of the first and the second terms on the right hand side of (5.4) follows in the same way as for the terms in (5.3). By Lemma 2.3 we obtain strong convergence to zero in $L_{2p}(Q_T)$ of the sequences

$$D^{\alpha'} \overline{G_n}(x,t),\ D^{\alpha''} v_0^{(n)}(x,t) \quad \text{for} \quad |\alpha'| = |\alpha''| = 1,$$

and this guarantees the strong convergence in $L_p(Q_T)$ of the last terms on the right hand side of (5.4). The proof is complete.

Proof of Theorem 3.1. Boundedness of the operator A defined by (3.11) follows from (3.7), (3.11). Using F_3, F_4 and the embedding theorem it is simple to prove the continuity of A. Further we will show that A satisfies the condition $(S)_+$ on $W_p^{(4),0}(Q_T)$.

Let $u_n(x) \in W_p^{(4),0}(Q_T)$ be a sequence such that $u_n(x) \rightharpoonup u_0(x)$ and

$$\overline{\lim_{n\to\infty}} \langle Au_n, u_n - u_0 \rangle \leq 0. \tag{5.5}$$

From (3.8), (3.11) we have

$$\langle Au_n, u_n - u_0 \rangle =$$

$$= \sum_{2s+|\alpha|\leq 2} \int_{Q_T} \psi_p \left\{ \left(\frac{\partial}{\partial t}\right)^s D^\alpha \left[\frac{\partial u_n}{\partial t} - F(x,t,u_n,D^1 u_n, D^2 u_n) - f\right]\right\} \cdot \tag{5.6}$$

$$\cdot \left(\frac{\partial}{\partial t}\right)^s D^\alpha \left[\frac{\partial(u_n - u_0)}{\partial t} - L(u_n)(u_n - u_0)\right] dxdt,$$

where the operator $L(u)$ is defined by (3.4).

From the equality

$$F_n(x,t) - F_0(x,t) = \int_0^1 L(su_n + (1-s)u_0)(u_n - u_0) ds$$

with $F_n(x,t) = F(x,t,u_n,D^1 u_n, D^2 u_n)$, $F_0(x,t) = F(x,t,u_0,D^1 u_0, D^2 u_0)$, and from Lemma 5.1 we obtain that the sequences

$$F_n(x,t) - F_0(x,t) - L(u_0)(u_n - u_0),$$

$$L(u_n)(u_n - u_0) - L(u_0)(u_n - u_0)$$

converge strongly to zero in $W_p^{(2)}(Q_T)$. This convergence and (5.5), (5.6) imply the inequality

$$\overline{\lim_{n\to\infty}} \sum_{2s+|\alpha|\le 2} \int_{Q_T} \psi_p \left\{ \left(\frac{\partial}{\partial t}\right)^s D^\alpha \left[\frac{\partial u_n}{\partial t} - F_0 - L(u_0)(u_n - u_0) - f \right] \right\} \cdot \left(\frac{\partial}{\partial t}\right)^s D^\alpha \left[\frac{\partial (u_n - u_0)}{\partial t} - L(u_0)(u_n - u_0) \right] dxdt \le 0. \tag{5.7}$$

From the weak convergence of $u_n(x,t)$ to $u_0(x,t)$ we have

$$\lim_{n\to\infty} \sum_{2s+|\alpha|\le 2} \int_{Q_T} \psi_p \left\{ \left(\frac{\partial}{\partial t}\right)^s D^\alpha \left[\frac{\partial u_0}{\partial t} - F_0(x,t) - f(x,t) \right] \right\} \cdot \left(\frac{\partial}{\partial t}\right)^s D^\alpha \left[\frac{\partial}{\partial t}(u_n - u_0) - L(u_0)(u_n - u_0) \right] dxdt = 0. \tag{5.8}$$

Note that for arbitrary $s_1, s_2 \in R^1$ we have

$$[\psi_p(s_1) - \psi_p(s_2)](s_1 - s_2) \ge C_p |s_1 - s_2|^p \tag{5.9}$$

with some positive constant C_p. From (5.7) - (5.9) we obtain

$$\overline{\lim_{n\to\infty}} \sum_{2s+|\alpha|\le 2} \int_{Q_T} \left| \left(\frac{\partial}{\partial t}\right)^s D^\alpha \left[\frac{\partial (u_n - u_0)}{\partial t} - L(u_0)(u_n - u_0) \right] \right|^p dxdt \le 0,$$

and consequently we have

$$\overline{\lim_{n\to\infty}} \left\| \frac{\partial (u_n - u_0)}{\partial t} - L(u_0)(u_n - u_0) \right\|_{p,Q_T}^{(2)} = 0. \tag{5.10}$$

By using the a priori estimate (4.4) we obtain

$$\|u_n - u_0\|_{p,Q_T}^{(4)} \le K_2 \left\| \frac{\partial (u_n - u_0)}{\partial t} - L(u_0)(u_n - u_0) \right\|_{p,Q_T}^{(2)},$$

and the strong convergence of the sequence $u_n(x,t)$ in $W_p^{(4)}(Q_T)$ follows from (5.10). The proof is complete.

Proof of Theorem 3.2. From the definition of the operator E (3.8) it follows immediately that $E(u, 0) = 0$ for arbitrary $u \in \mathring{W}_p^{(4),0}(Q_T)$. Then from (3.2),

(3.11) we obtain that the solution of the problem (3.2) is a solution of the operator equation $Au = 0$.

Let $u_0(x,t) \in \overset{\circ}{W}{}_p^{(4),0}(Q_T)$ be a solution of the equation $Au = 0$. By using (3.7) we obtain that the function

$$h_0(x,t) = \frac{\partial u_0}{\partial t} - F(x,t,u_0,D^1u_0,D^2u_0) - f(x,t)$$

belongs to $W_p^{(2),0}(Q_T)$. From Theorem 5.1 it follows that there exists a function $\varphi_0(x.t) \in \overset{\circ}{W}{}_p^{(4),0}(Q_T)$ such that

$$\frac{\partial \varphi_0(x,t)}{\partial t} - L(u_0)\,\varphi_0(x,t) = h_0(x,t).$$

We have the equality

$$0 = \langle Au_0, \varphi_0\rangle = \langle E(u_0,h_0),\varphi_0\rangle = \int\limits_{Q_T} \sum_{2s+|\alpha|\leq 2} \psi_p\left\{\left(\frac{\partial}{\partial t}\right)^s D^\alpha h_0(x,t)\right\}\cdot$$

$$\cdot\left(\frac{\partial}{\partial t}\right)^s D^\alpha h_0(x,t)dxdt = \int\limits_{Q_T} \sum_{2s+|\alpha|\leq 2} \left\|\left(\frac{\partial}{\partial t}\right)^s D^\alpha h_0(x,t)\right\|^p dx\,dt,$$

from which it follows that $h_0(x,t) = 0$ and consequently $u_0(x,t)$ is the solution of the problem (3.2). The proof is complete.

amstex

§6. Solvability of nonlinear parabolic problems

We will start with the uniqueness theorem and the proof of local solvability of the problem (2.17) - (2.19).

Proof of Theorem 3.3. We assume for a contradiction that there exist two solutions $u_1(x,t), u_2(x,t)$ of the problem (3.2). Then the function $u(x,t) = u_1(x,t) - u_2(x,t)$ is a solution of the linear parabolic problem (4.1), (4.2) with $f(x,t) \equiv 0$ and

$$a_\alpha(x,t) = \int_0^1 \left[\widetilde{F}_\alpha(su_1 + (1-s)u_2)\right](x,t)\,dx\,dt, \tag{6.1}$$

where $\widetilde{F}_\alpha$ is understood as in (2.31). The coefficients $a_\alpha(x,t)$ defined by (6.1) satisfy the conditions a_1, a_2 of §5 and we obtain the equality $u_1(x,t) = u_2(x,t)$ by Theorem 4.1. The uniqueness of the solution of the initial boundary value problem (3.2) is proved.

Proof of Theorem 3.4. Local solvability of the problem (2.17) - (2.19) will follow from Theorems 1.4, 1.5. For the linear operator $L(0)$ defined by (3.4) with $u(x,t) \equiv 0$ and for $0 < T \leq 1$ we have from (3.8), (4.4)

$$\begin{aligned} \langle E(0, \frac{\partial u}{\partial t} - L(0)u), u\rangle &= \sum_{2s+|\alpha|\leq 2} \int_{Q_T} \left| \left(\frac{\partial}{\partial t}\right)^s D^\alpha \left[\frac{\partial u}{\partial t} - L(0)u\right]\right|^p dxdt = \\ &= \left\{ \left\| \frac{\partial u}{\partial t} - L(0)u \right\|^{(2)}_{p,Q_T} \right\}^p \geq C^{(0)} \{ \|u\|^{(4)}_{p,Q_T} \}^p, \end{aligned} \tag{6.2}$$

where $u(x,t) \in \overset{\circ}{W}{}_p^{(4),0}(Q_T)$ and $C^{(0)}$ is some positive constant independent of $u(x,t), T$.

From the reduction of the problem (2.17) - (2.19) to the problem with zero initial and boundary data it follows that it suffices to prove the local solvability of (3.2). We will prove the estimate

$$|\langle Au, u\rangle - \langle E(0, \frac{\partial u}{\partial t} - L(0)u), u\rangle| \leq \frac{1}{2} C^{(0)} R^p \tag{6.3}$$

if

$$\|f(x,t)\|^{(2)}_{p,Q_T} \leq K, \quad \|u(x,t)\|^{(4)}_{p,Q_T} = R, \tag{6.4}$$

for a given number K and a suitable choice of R and T_0.

Let ε be an arbitrary positive number. Then by inequalities (4.14) - (4.17) it is possible to choose a positive number $T_0(R,\varepsilon)$ such that for $u(x,t) \in \overset{\circ}{W}{}_p^{(4),0}(Q_T)$

$$0 < T \leq T_0(R,\varepsilon), \quad \|u\|^{(4)}_{p,Q_T} \leq R, \tag{6.5}$$

the following inequalities

$$\begin{aligned} &\sum_{2s+|\gamma|\leq 2} \left\| \left(\frac{\partial}{\partial t}\right)^s D^\gamma u(x,t) \right\|^{(\delta)}_{Q_T} \leq \varepsilon \quad \text{for } 0 \leq \delta < 2 - \frac{n+2}{p} \\ &\sum_{2s+|\gamma|=3} \left\| \left(\frac{\partial}{\partial t}\right)^s D^\gamma u(x,t) \right\|_{2p,Q_T} \leq \varepsilon \end{aligned} \tag{6.6}$$

are valid. For $\varepsilon \leq 1$ by using the inequalities (6.6) we obtain

$$\|L(u)u\|^{(2)}_{p,Q_T} \leq C^{(1)} \|u\|^{(4)}_{p,Q_T}, \quad \|F(x,t,u,D^1u,D^2u)\|^{(2)}_{p,Q_T} \leq C^{(1)} \|u\|^{(4)}_{p,Q_T} \tag{6.7}$$

for u, T satisfying the conditions (6.5) with a positive constant $C^{(1)}$ independent of u, R, ε. The proof of the estimates (6.7) is analogous to that of (3.5), (3.7). Analogously it is possible to prove the inequalities

$$\begin{aligned} \|L(u)u - L(0)u\|^{(2)}_{p,Q_T} &\leq C(\varepsilon) \|u\|^{(4)}_{p,Q_T} \\ \|F(x,t,u,D^1u,D^2u) - L(0)u\|^{(2)}_{p,Q_T} &\leq C(\varepsilon) \|u\|^{(4)}_{p,Q_T} \end{aligned} \tag{6.8}$$

for u, R satisfying the conditions (6.5) where $C(\varepsilon)$ tends to zero if ε tends to zero.

We evaluate the left hand side of (6.3) by using (6,6), (6.7) and $\widetilde{F}(u) = $
$= F(x,t,u,D^1u,D^2u)$:

$$\begin{aligned} &|\langle Au,u\rangle - \langle E(0,\frac{\partial u}{\partial t} - L(0)u,u\rangle| \le \\ &\le |\langle E(u,\frac{\partial u}{\partial t} - \widetilde{F}(u) - f) - E(0,\frac{\partial u}{\partial t} - \widetilde{F}(u) - f,u\rangle| + \\ &+|\langle E(0,\frac{\partial u}{\partial t} - \widetilde{F}(u) - f) - E(0,\frac{\partial u}{\partial t} - \widetilde{F}u),u\rangle| + \\ &+|\langle E(0,\frac{\partial u}{\partial t} - \widetilde{F}(u)) - E(0,\frac{\partial u}{\partial t} - L(0)u),u\rangle| \le \\ &\le C^{(2)}\{[R^{p-1} + K^{p-1}]\,C(\varepsilon)\,R + KR^{p-1}\}. \end{aligned} \tag{6.9}$$

The last inequality was obtained from the assumptions (6.4), (6.5), $C(\varepsilon)$ in (6.9) is the same as in (6.8), $C^{(2)}$ is a positive constant independent of $u(x,t), R, \varepsilon, T_0(R,\varepsilon)$.

By the Young inequality we have that the right-hand side of (6.9) is less than

$$\frac{C^{(0)}}{4}R^p + C^{(3)}K^p + C^{(2)}C(\varepsilon)R^p.$$

If the number K is fixed we define R by the equality

$$R = \left[\frac{8}{C^{(0)}}C^{(3)}\right]^{\frac{1}{p}} K$$

and choose ε satisfying condition $8C^{(2)}C(\varepsilon) < C^{(0)}$. For such choice of R, ε and $T_0 = T_0(R,\varepsilon)$ we have (6.3) for $0 < t \le T_0$ and $u(x,t)$ satisfying (6.4). Consequently we obtain from (6.2), (6.3) the inequality

$$\langle Au,u\rangle \ge \frac{C^{(0)}}{2}R^p \quad \text{for} \quad \|u(x,t)\|^{(4)}_{p,Q_T} = R. \tag{6.10}$$

The existence of a solution u of the equation $Au = 0$ in the ball $B(0,R)$ follows from last inequality and Theorems 1.4, 1.5. By Theorem 3.3, $u(x,t)$ is a solution of (3.2) and the proof of local solvability is complete.

Proof of Theorem 3.5. Analogously to the proof of Theorem 3.2 we reduce the problem for the equation (3.16) with zero boundary and initial conditions to the operator equation

$$A_\lambda u = 0, \quad u \in \overset{\circ}{W}{}^{(4),0}_p(Q_T), \tag{6.11}$$

with $\lambda \in [0,1]$. The problem (3.2) is equivalent to the equation (6.11) for $\lambda = 1$.

According to Theorem 3.1 for an arbitrary number $\lambda \in [0,1]$, A_λ is a bounded continuous operator of class $(S)_+$. It is possible to verify that for the ball

$$B(R+1) = \left\{ u(x,t) \in \overset{\circ}{W}{}_p^{(4),0}(Q_T) : \|u\|_{p,Q_T}^{(4)} < R+1 \right\} \tag{6.12}$$

all conditions of Definition 1.5 about homotopy of the operators A_0 and A_1 are satisfied. The number R in (6.12) is defined in Theorem 3.5. We verify that the condition h_3 is fulfilled analogously to the proof of Theorem 3.3.

For the proof of the solvability of (3.2) is suffices to show that

$$\mathrm{Deg}(A_0, \overline{B(R+1)}, 0) \neq 0. \tag{6.13}$$

Indeed by the inequality (6.13) the existence of the solution of the equation (6.11) follows from Theorems 1.3, 1.4.

For a given positive number ε it is possible to choose a number $\rho > 0$ such that for $u \in \partial B(\rho)$ the inequalities (6.7), (6.8) are valid. In the same way as in the proof of the inequality (6.10) we obtain $\langle Au, u \rangle \geq 0$ for $u \in \partial B(\rho)$ and by Theorem 1.6 we have

$$\mathrm{Deg}(A_0, \overline{B(\rho)}, 0) = 1. \tag{6.14}$$

From the additivity of the degree with respect to the domain (an analogue of the property d_2) for the degree of a finite dimensional mapping) and from the uniqueness of the solution of the equation $A_0 u = 0$ we obtain from (6.14)

$$\mathrm{Deg}(A_0, \overline{B(R+1)}, 0) = 1. \tag{6.15}$$

Therefore the inequality (6.13) is proved and we finished the proof of Theorem 3.5.

Remark 6.1. Analogously to the proof of (6.15) it is possible to establish the following property of the degree of the operator A defined by (3.11). Let D be a bounded open set in $\overset{\circ}{W}{}_p^{(4),0}(Q_T)$ such that $Au \neq 0$ for $u \in \partial D$. Then

$$\mathrm{Deg}(A, \overline{D}, 0) = 1$$

if and only if there exists a solution u of the equation $Au = 0$ in D. In the opposite case $Deg(A, \overline{D}, 0) = 0$.

Finally we formulate a result about the invariance of a domain and the applications of Theorem 1.7 to the problem (3.2) (analogous applications can be formulated for the problem (2.17) - (2.19)).

Theorem 6.1. *Assume that the hypotheses of Theorem 3.1 are satisfied, D is an open set in $\overset{\circ}{W}{}_p^{(4),0}(Q_T)$, and let $R(D)$ be a set of all functions $f(x,t) \in W_p^{(2),0}(Q_T)$ such that the problem (3.2) has a solution $u(x,t) \in D$ for $f(x,t) \in R(D)$. Then the set $R(D)$ is open in $W_p^{(2),0}(Q_T)$.*

The assertion of Theorem 6.1 follows from Theorem 7.4, Chapter 2 [10].

For the construction of the approximate solution for the problem (3.2) we introduce some complete system $\{v_i(x,t)\}$, $i = 1.2....$, in the space $\overset{\circ}{W}{}_p^{(4),0}(Q_T)$ such that for an arbitrary n the functions $v_1(x,t), ..., v_n(x,t)$ are linearly independent. We will call an approximate solution for the problem (3.2) a function $u_n(x,t) = C_1^{(n)} v_1(x,t) + ... + + C_n^{(n)} v_n(x,t)$ such that the following equalities

$$\sum_{2s+|\alpha|\leq 2} \int_{Q_T} \psi_p \left\{ \left(\frac{\partial}{\partial t}\right)^s D^\alpha \left[\frac{\partial u_n}{\partial t} - F(x,t,u_n,D^1 u_n, D^2 u_n) - f(x,t)\right]\right\} \cdot \left(\frac{\partial}{\partial t}\right)^s D^\alpha \left[\frac{\partial v_i}{\partial t} - L(u_n) v_i\right] dx dt = 0, \quad i = 1, ..., n \tag{6.16}$$

are valid. In (6.16) we use the same ingredients as in (3.8).

Theorem 6.2. *Assume that the hypotheses of Theorem 3.1 are satisfied. Suppose that D is a bounded open set in $\overset{\circ}{W}{}_p^{(4),0}(Q_T)$ and the problem (3.2) has a solution $u_0(x,t)$ in D. Then for sufficiently large n there exist approximate solutions $u_n(x,t)$ of the problem (3.2) defined by (6.16), and the sequence $u_n(x,t)$ strongly converges to $u_0(x,t)$ in $\overset{\circ}{W}{}_p^{(4),0}(Q_T)$.*

The assertion of Theorem 6.2 follows immediately from Theorems 1.7, 3.1 - 3.3 and Remark 6.1.

REFERENCES

1. P.ACQUISTAPACE, B.TERRENI, Fully nonlinear parabolic systems, *in: Proceedings of the Conference "Nonlinear Elliptic and Parabolic Problems", Nancy,* March 14-19 (1988).
2. S.I.HUDJAEV, The first boundary value problem fon non-linear parabolic equations, *Dokl. Akad. Nauk SSSR* **149** (1963), 535-538.
3. V.P.IL'IN, The properties of some classes of differentiable functions of several variables defined in an n-dimensional region, *Trudy Math. Inst. Steklov* **66** (1962), 227-363.
4. S.N.KRUZHKOV, A.CASTRO, M.LOPES, Schauder type estimates and theorems on the existence of the solution of fundamental problem for linear and nonlinear parabolic equations, *Dokl. Akad. Nauk SSSR* **220** (1975), 277-280.
5. N.V.KRYLOV, Nonlinear elliptic and parabolic equations of the second order, *"Nauka", Moscow,* 1985.
6. O.A.LADYZHENSKAJA, V.A.SOLONNIKOV, N.N.URAL'TSEVA, Linear and quasilinear equations of parabolic type, *"Nauka", Moscow,* 1967.

7. A.LUNARDI, On a class of fully nonlinear parabolic equation, *Comm. Part. Dif. Equat.* **16(1)** (1991), 145-172.
8. L.NIRENBERG, On elliptic partial differential equations, *Ann. Scuola Norm. Sup. Pisa* **(3)13** (1959), 115-162.
9. I.V.SKRYPNIK, Nonlinear higher order elliptic equations, *"Naukova dumka", Kiev*, 1973.
10. I.V.SKRYPNIK, Methods for analysis of nonlinear elliptic boundary value problems, *Trans. Math. Monographs, V.139, Providence, A.M.S.*, 1994.
11. V.A.SOLONNIKOV, A priori estimates for second order parabolic equations, *Trudy Math. Inst. Steklov* **70** (1964), 133-212.
12. N.N.SOPOLOV, The first boundary value problem for nonlinear parabolic equations of any order, *C.R. Acad. Bulgare Sci.* **23** (1970), 899-902.

I. V. Skrypnik
Institut of Applied Mathematics and Mechanics
Ukrainian Academy of Sciences
74, Roza Luxemburg str.
Donetsk – 114
Ukraine